electronics five

Hayden Electronics One-Seven Series

Harry Mileaf, Editor-in-Chief

electronics one — Electronic Signals ☐ Amplitude Modulation ☐ Frequency Modulation ☐ Phase Modulation ☐ Pulse Modulation ☐ Side-Band Modulation ☐ Multiplexing ☐ Television Signals ☐ Navigation Signals ☐ Facsimile ☐ Heterodyning ☐ Harmonics ☐ Waveshaping

electronics two — Electronic Building Blocks ☐ Basic Stages ☐ The Power Supply ☐ Transmitters ☐ Receivers ☐ UHF ☐ Telemetry ☐ Television ☐ Radar and Sonar ☐ RDF ☐ Radio Navigation ☐ Radio Control

electronics three — Electron Tubes ☐ Diodes ☐ Triodes ☐ Tetrodes ☐ Pentodes ☐ Multielement Tubes ☐ Gas-Filled Tubes ☐ Phototubes ☐ Electron-Ray Indicators ☐ Cathode-Ray Tubes ☐ UHF and Microwave Tubes ☐ Magnetrons ☐ Klystrons ☐ The Traveling-Wave Tube

electronics four — Semiconductors ☐ P-N Diodes ☐ Avalanche Diodes ☐ Switching Diodes ☐ Zener Diodes ☐ Tunnel Diodes ☐ N-P-N and P-N-P Transistors ☐ Tetrode Transistors ☐ Field Effect, Surface Barrier, Unijunction, SCR Transistors ☐ Photodevices

electronics five — Power Supplies ☐ Rectifiers ☐ Filters ☐ Voltage Multipliers ☐ Regulation ☐ Amplifier Circuits ☐ A-F, R-F, and I-F Amplifiers ☐ Video Amplifiers ☐ Phase Splitters ☐ Follower Amplifiers ☐ Push-Pull Amplifiers ☐ Limiters

electronics six — Oscillators ☐ Sinusoidal and Nonsinusoidal Oscillators ☐ Relaxation Oscillators ☐ Magnetron Oscillators ☐ Klystron Oscillators ☐ Crystal Oscillators ☐ Modulators ☐ Mixers and Converters ☐ Detectors and Demodulators ☐ Discriminators

electronics seven — Auxiliary Circuits ☐ AVC, AGC, and AFC Circuits ☐ Limiter and Clamping Circuits ☐ Separator, Counter, and Gating Circuits ☐ Time Delay Circuits ☐ Radio Transmission ☐ Antennas ☐ Radiation Patterns ☐ R-F Transmission Lines

electronics five

HARRY MILEAF EDITOR-IN-CHIEF

5

HAYDEN BOOK COMPANY, INC.
Rochelle Park, New Jersey

Library of Congress Catalog Card Number 66-23489

Printed in the United States of America

Second Printing 1967

preface

This volume is one of a series designed specifically to teach electronics. The series is logically organized to fit the learning process. Each volume covers a given area of knowledge, which in itself is complete, but also prepares the student for the ensuing volumes. Within each volume, the topics are taught in incremental steps and each topic treatment prepares the student for the next topic. Only *one* discrete topic or concept is examined on a page, and *each* page carries an illustration that graphically depicts the topic being covered. As a result of this treatment, neither the text nor the illustrations are relied on solely as a teaching medium for any given topic. Both are given for *every* topic, so that the illustrations not only complement but reinforce the text. In addition, to further aid the student in retaining what he has learned, the important points are summarized in text form on the illustration. This unique treatment allows the book to be used as a convenient review text. Color is used not for decorative purposes, but to accent important points and make the illustrations meaningful.

In keeping with good teaching practice, all technical terms are defined at their point of introduction so that the student can proceed with confidence. And, to facilitate matters for both the student and the teacher, key words for each topic are made conspicuous by the use of italics. Major points covered in prior topics are often reiterated in later topics for purposes of retention. This allows not only the smooth transition from topic to topic, but the reinforcement of prior knowledge just before the declining point of one's memory curve. At the end of each group of topics comprising a lesson, a summary of the facts is given, together with an appropriate set of review questions, so that the student himself can determine how well he is learning as he proceeds through the book.

Much of the credit for the development of this series belongs to various members of the excellent team of authors, editors, and technical consultants assembled by the publisher. Special acknowledgment of the contributions of the following individuals is most appropriate: Frank T. Egan, Jack Greenfield, and Warren W. Yates, principal contributors; Peter J. Zurita, S. William Cook, Jr., Steven Barbash, Solomon Flam, and A. Victor Schwarz, of the publisher's staff; Paul J. Barotta, Director of the Union Technical Institute; Albert J. Marcarelli, Technical Director of the Connecticut School of Electronics; Howard Bierman, Editor of *Electronic Design;* E. E. Grazda, Editorial Director of *Electronic Design;* and Irving Lopatin, Editorial Director of the Hayden Book Companies.

HARRY MILEAF
Editor-in-Chief

contents

power supplies

The electron tubes and transistors used in electronic equipment require definite operating voltages for proper operation. A power supply is normally used to furnish these voltages at the necessary current ratings. Except for the filament voltage of tubes, operating voltages must be nearly *pure dc*. Some equipment contains batteries to supply these d-c voltages. In most equipment, though, some form of *electronic power supply* is used. These power supplies convert a-c input power, usually from a commercial source, into one or more d-c voltages. In equipment that uses electron tubes, the power supply also furnishes the low-voltage, relatively high-current a-c power required by the tube filaments.

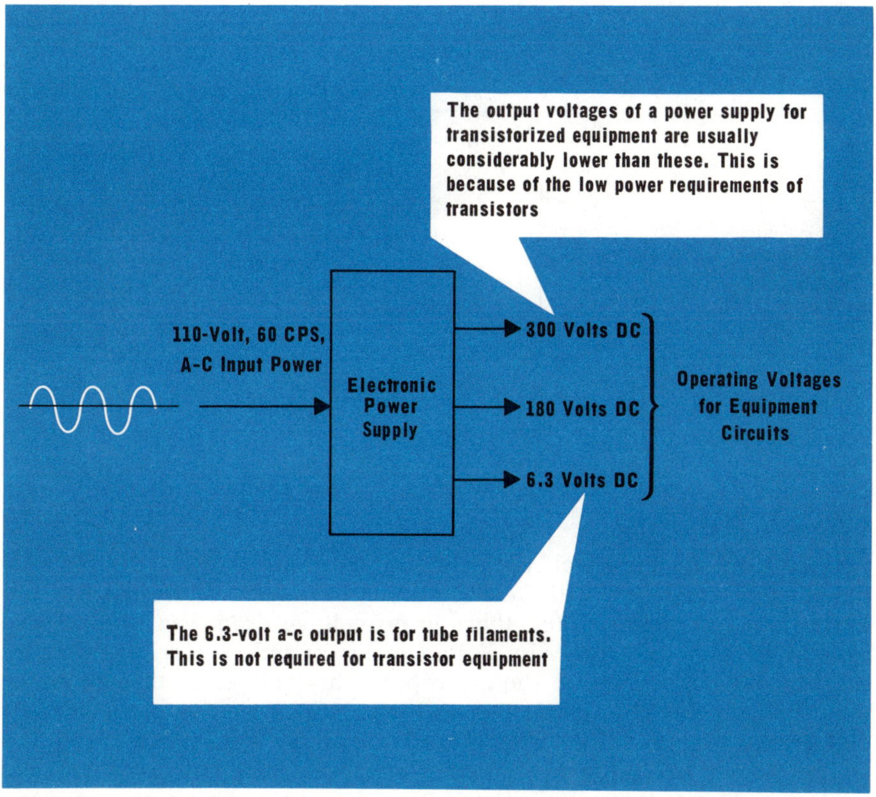

To accomplish its function, a power supply: (1) *rectifies* the sinusoidal a-c power input; and (2) smooths, or *filters*, the resulting fluctuating dc. In addition, many power supplies contain a transformer to step up the a-c input voltage before rectification and filtering take place, and also contain regulator circuits to keep the d-c voltages from fluctuating.

the basic half-wave rectifier

Rectification, you recall, is the removal of *one-half* of a sinusoidal or other symmetrical wave. One of the most widely used devices for accomplishing rectification is the electron-tube or semiconductor diode. This is because of the diode's *unidirectional* conduction characteristics. The diode only conducts in one direction. When a sinusoidal voltage is applied between the plate and cathode, this condition exists only during alternate half cycles.

A basic diode rectifier circuit is shown. In the electron tube, the positive half cycles of the a-c input voltage cause the plate of the diode to be positive with respect to the cathode. As a result, the tube conducts and current flows through resistor R. The voltage drop produced by this current has the same waveshape as the half cycle of a-c input voltage that causes the diode to conduct.

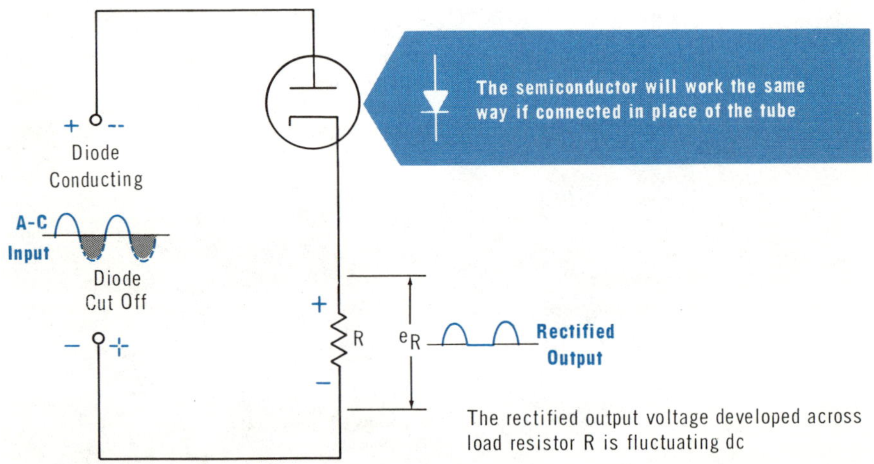

The rectified output voltage developed across load resistor R is fluctuating dc

During the negative half cycles of the a-c input voltage, the plate of the diode is negative with respect to the cathode. Consequently, the diode does not conduct, and there is no voltage drop developed across resistor R. The voltage across R is a series of pulses corresponding to the positive half cycles of the a-c input. This voltage is the output of the rectifier. Such a rectifier is called a *half-wave rectifier,* since it delivers an output during only half of the input voltage cycle.

The output voltage from a rectifier can be increased by applying the a-c input voltage through a *step-up* transformer. Such an arrangement is shown in A on the next page. Although the available a-c source voltage is 115 volts, a transformer with a step-up turns ratio increases the voltage to 450 volts. In this case, this is done before the voltage is applied between the plate and cathode of the diode. Such a transformer is called a power transformer, since it normally handles a considerable amount of electrical power.

other half-wave rectifiers

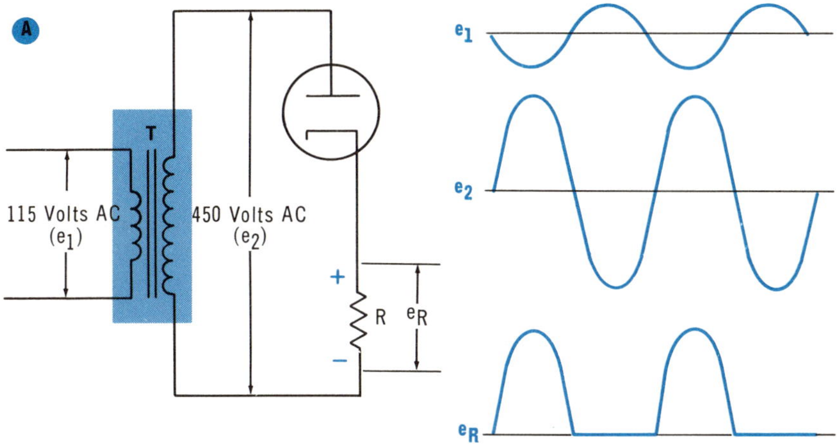

A half-wave rectifier circuit using a semiconductor rectifier is shown in B. This circuit is the same as the tube circuit shown in A, except that the tube diode has been replaced by a semiconductor diode. The semiconductor diode has a low forward resistance, and, therefore, conducts heavily when the applied voltage has the polarity shown. When the voltage polarity reverses, the diode conducts very little. This reverse current is so small that for practical purposes it can be neglected.

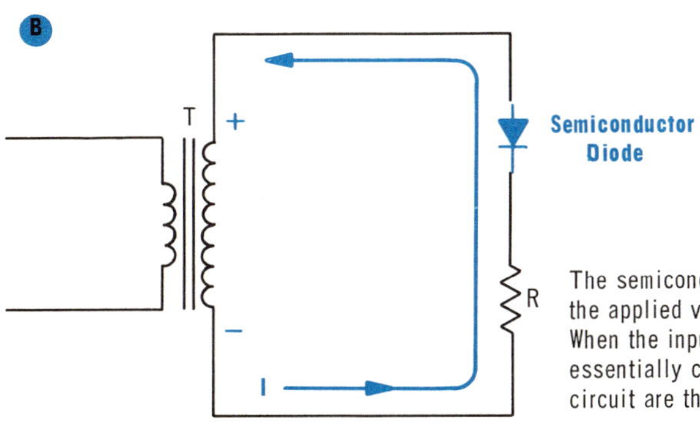

The semiconductor diode conducts only when the applied voltage has the polarity shown. When the input polarity reverses, the diode essentially cuts off. The waveforms in this circuit are the same as those in A

In other words, the diode will conduct electron current in the *forward* direction, which is opposite to the direction in which the arrow portion of the diode symbol points. The diode will not conduct in the direction of the arrow, which is called the *reverse* direction. The arrow portion of the diode symbol thus corresponds to the plate of the vacuum-tube diode, and is sometimes called the *anode*. The arrow is said to point in the direction of *high* resistance.

characteristics of half-wave rectifiers

A half-wave rectifier delivers an output during only half of the a-c input cycle. The average output voltage is therefore not equal to 0.637 times the peak voltage (e_{PEAK}), as is the case for a full 360-degree sinusoidal wave. Instead, the average voltage is equal to one-half of 0.637 e_{PEAK}, or 0.318 e_{PEAK}.

The half-wave rectifier draws power from the input power source during only half of the input cycle. It is therefore limited in the amount of *current* it can supply to the loads that it powers. For this reason, the half-wave rectifier is usually used only in applications that require a relatively small current drain.

The average output voltage of a half-wave rectifier is only 0.318 times the peak input voltage

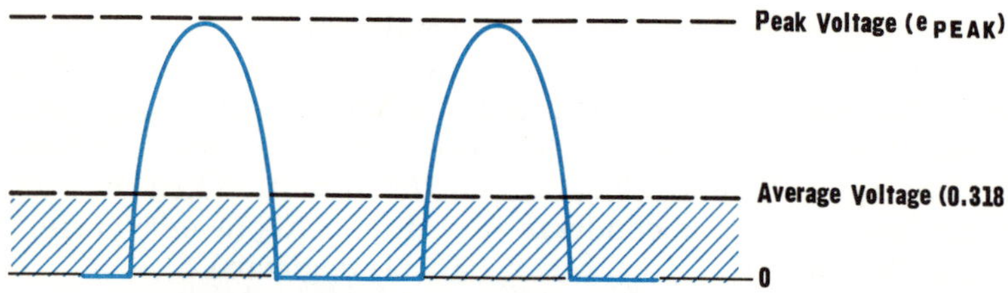

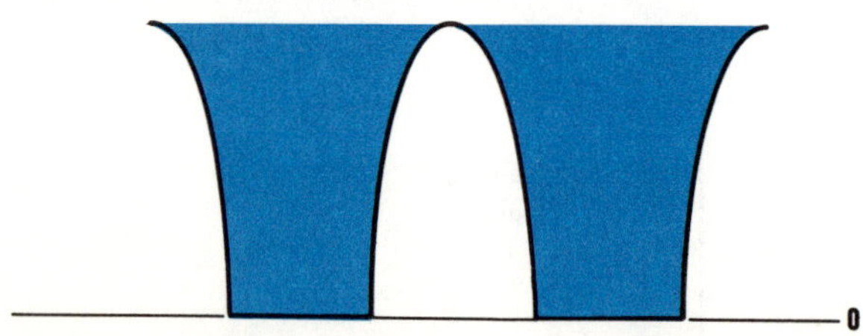

With the half-wave rectifier, the actual voltage is zero a great deal of the time. As a d-c voltage, it leaves much to be desired

Another disadvantage of the half-wave rectifier is that the current flow in the power transformer secondary is always in the same direction. This causes a considerable amount of *core saturation* in the transformer iron core and reduces the efficiency of the transformer.

Probably the biggest disadvantage of the half-wave rectifier is that its output is far from a *perfect* d-c voltage. This is because the individual output pulses are relatively widely spaced with periods of no output voltage between them. Where a steady d-c output is required, therefore, the half-wave rectifier is unsuitable.

Of course, the polarity of the output voltage can be changed by reversing the diode connection, so that the other half of the sine wave will be rectified.

the full-wave rectifier

The disadvantages of the half-wave rectifier can be overcome by using what is called a *full-wave rectifier*. In this type of rectifier, *two* diodes are connected so that each conducts during alternate half cycles of the a-c input. The two diodes have a *common load* and the current flow through this load is always in the same direction.

The full-wave rectifier uses two diodes and a common-load resistor, R. Either vacuum-tube diodes or semiconductor diodes can be used

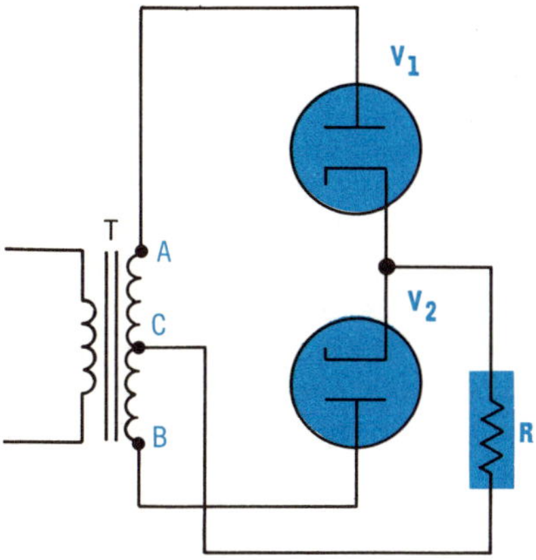

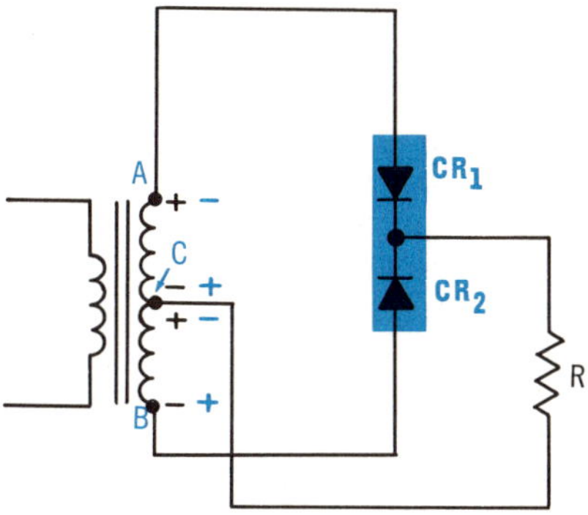

Point C is always at the same potential, but relative to A and B, it is at two different potentials. When A is positive with respect to B, C is negative with respect to A and positive with respect to B. The opposite is true when A is negative with respect to B

A basic full-wave rectifier circuit is shown. The circuit uses two diodes, with the anodes of the diodes connected to opposite ends of the transformer secondary. The common-cathode resistor, R, is connected between the two cathodes and a centertap on the secondary winding of the transformer. When an a-c input is applied to the transformer primary, it appears as a stepped-up voltage across the entire transformer secondary (points A and B). Since point C is the *electrical center* of the secondary, half of the induced voltage exists between points A and C, and half between B and C. These two voltages, however, are always 180 degrees out of phase, since they are measured with respect to the common reference point, C. The voltage between A and C is applied between the plate and cathode of diode V_1, and the voltage between B and C is applied to V_2.

Special full-wave rectifier tubes are made that have two diodes in one envelope. They can have either common or separate cathodes. Full-wave semiconductor diodes are also made as integral units.

operation of the full-wave rectifier

The two 180-degree, out-of-phase voltages developed across the center-tapped transformer secondary cause the two diodes of the full-wave rectifier to conduct on alternate half cycles of the a-c input. This is shown on the following page. During each negative half cycle of the a-c input, as shown in A, the end of the secondary winding connected to diode V_1 is positive and the end connected to V_2 is negative. Point C is therefore negative with respect to point A and positive with respect to point B. This means that the plate of V_1 is positive with respect to its cathode and the plate of V_2 is negative with respect to its cathode. As a result, diode V_1 conducts and diode V_2 is cut off. The current path for V_1 is from point C, through load resistor R, through V_1, and back to point A of the transformer. The voltage developed across resistor R has a waveshape similar to the half cycle of a-c input and a polarity as shown.

During each positive half cycle of the input, the voltage polarity across the transformer secondary *reverses*. Now, the plate of V_2 is positive with respect to its cathode and the plate of V_1 is negative with respect to its cathode. V_2 conducts and V_1 is cut off. The current path as V_2 conducts is shown in B on the following page. You can see that the polarity of the output pulse developed across the load resistor is the same as that produced by a negative half cycle of the a-c input. Each half cycle of the a-c input therefore produces output pulses across R all having the *same polarity*. This is shown in C.

The full-wave rectifier thus produces a rectified output for every half cycle of the a-c input, instead of only for alternate half cycles, as done by the half-wave rectifier. The average value of the output voltage is the same as that of a sine-wave voltage, or 0.637 times the peak voltage, and the rms value is 0.707 peak.

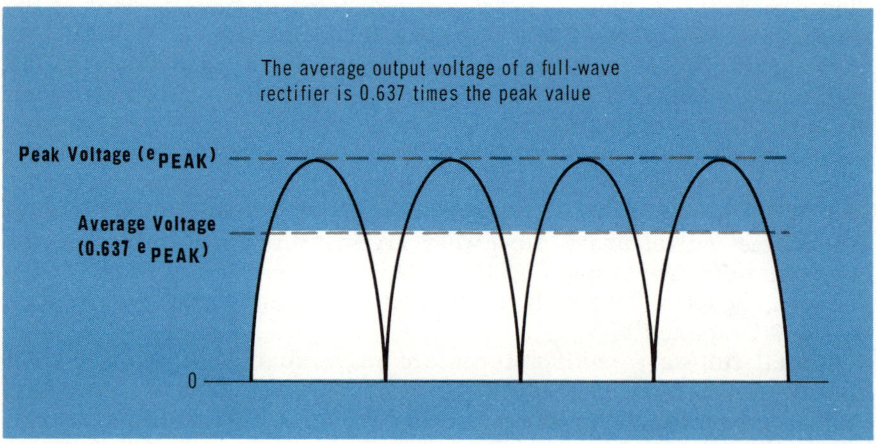

operation of the full-wave rectifier (cont.)

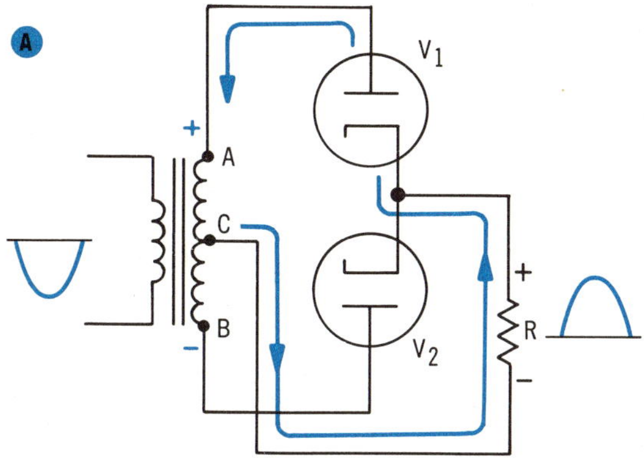

V_1 conducts for one polarity of voltage . . .

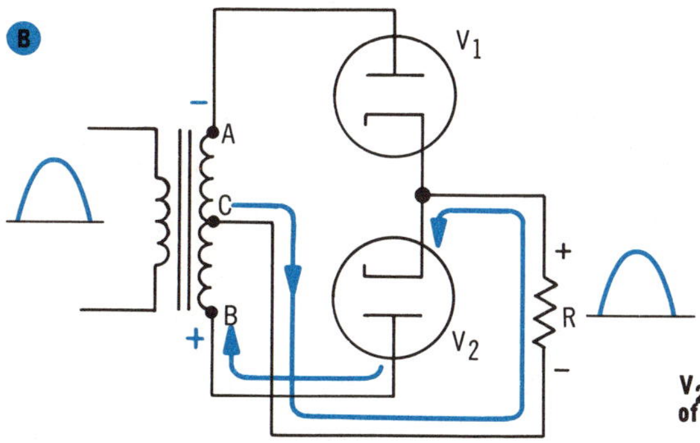

V_2 conducts for the other polarity of voltage

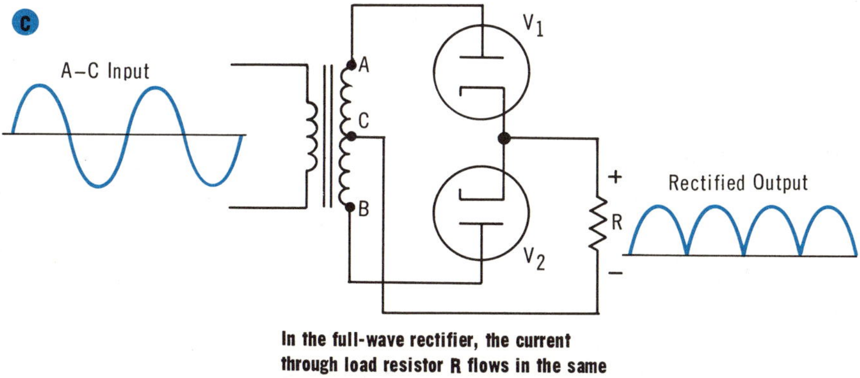

In the full-wave rectifier, the current through load resistor R flows in the same direction during each half cycle of the a-c input

the bridge rectifier

A special type of full-wave rectifier in which *four* diodes are used is shown. This full-wave rectifier is called a *bridge rectifier.* The a-c input is connected to two diagonally-opposite corners of the network and the rectified output is taken between the other two corners. During the half cycles of the a-c input when the top of the transformer secondary is positive and the bottom of the secondary is negative, the secondary voltage can be considered to be impressed across a voltage divider consisting of diode 1, load resistor R, and diode 3. Current therefore flows from the bottom of the secondary, through diodes 1 and 3 and resistor R, and back to the top of the transformer. This is shown in A on the next page.

A bridge rectifier employs four diodes with the a-c input and rectified output connections at opposite corners of the bridge network

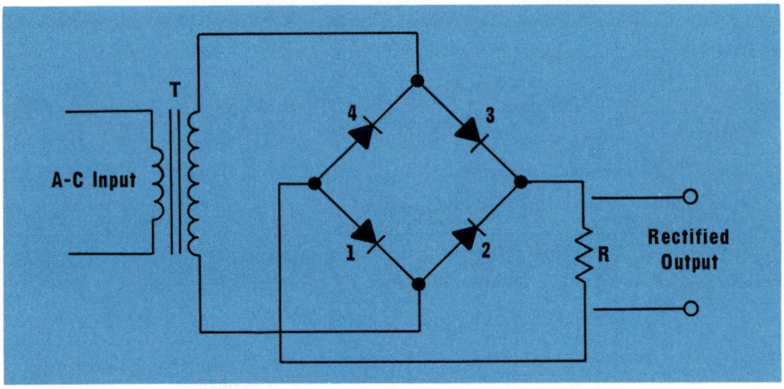

The rectifiers can be four separate tube or semiconductor diodes; some semiconductor bridge rectifiers, though, are made as one integral unit, with four rectifier sections

On the alternate half cycles of the a-c input, the bottom of the transformer secondary is positive and the top is negative. Diodes 1 and 3 cannot conduct, since the polarity of the voltage causes them to cut off. However, diodes 2 and 4 can now conduct. The current path, therefore, is from the top of the transformer, through diode 4, through the load resistor, through diode 2, and back to the bottom of the secondary. This is shown in B.

For both the positive and the negative half cycles of the a-c input, the direction of current flow through resistor R is the same. Therefore, the output voltage developed across R is a rectified version of the a-c input.

the bridge rectifier (cont.)

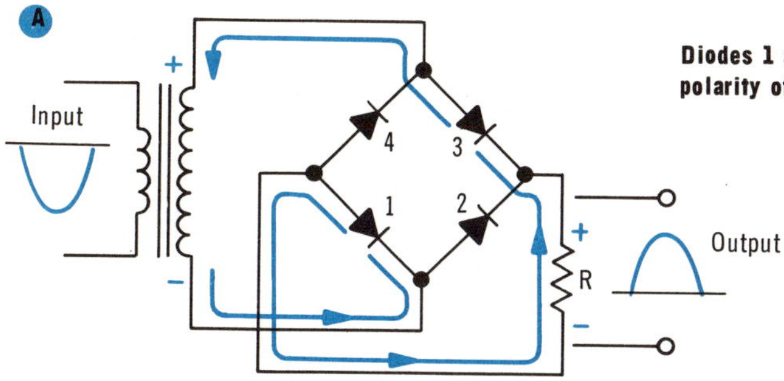

Diodes 1 and 3 conduct for one polarity of voltage . . .

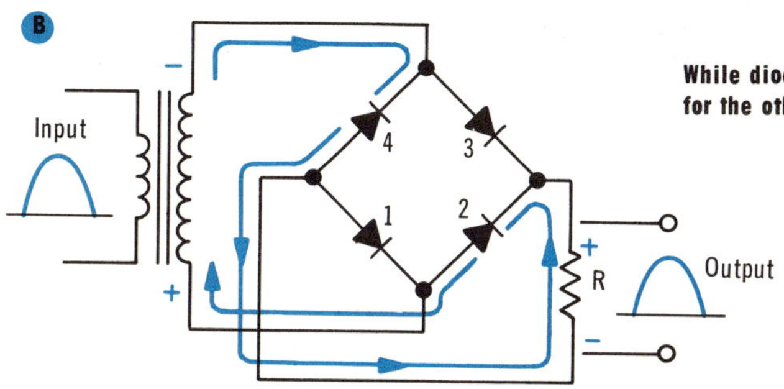

While diodes 2 and 4 conduct for the other polarity of voltage

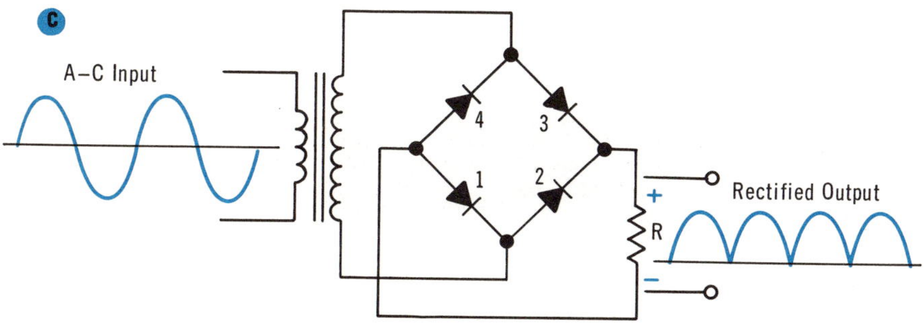

In the bridge rectifier, two of the four diodes conduct during the positive half cycles of the a-c input and the other two conduct during the negative half cycles. During both half cycles, the current flow through load resistor R is in the same direction

characteristics of the bridge rectifier

With a given power transformer, the bridge rectifier produces an output voltage nearly *twice* as large as that produced by a conventional full-wave rectifier. This is because the *entire* secondary voltage of the transformer is applied across each of the diode pairs. The instantaneous voltage developed across the load resistor is the full secondary voltage minus the relatively small voltage dropped across the two diodes. In the conventional full-wave rectifier, on the other hand, only *half* of the transformer secondary voltage is applied across each diode. So the voltage developed across the load resistor is only half the secondary voltage minus the drop across one diode.

If the power transformer and the a-c input are the same . . .

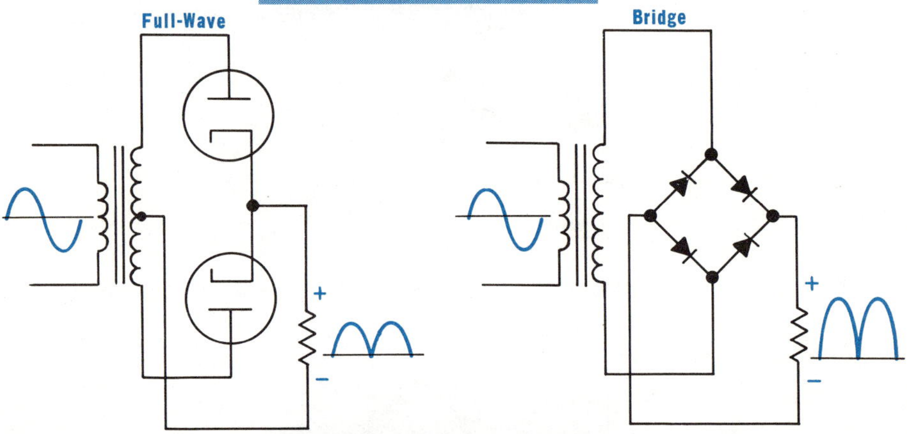

a bridge rectifier delivers an output nearly twice as large as that of a full-wave rectifier

Another advantage of the bridge rectifier is that, for a given output voltage, the *peak inverse voltage* across each diode is only *half* that which occurs in a conventional full-wave rectifier. You recall that the peak inverse voltage is the maximum voltage across a diode when its plate is negative with respect to its cathode. If this voltage exceeds the diode's rated value, breakdown occurs within the diode.

A disadvantage of the bridge rectifier is that a single-filament transformer cannot supply filament voltage for all four tubes when electron-tube diodes are used. Because of this necessity for multiple-filament transformers, tubes are not used frequently in bridge rectifier circuits.

output polarities

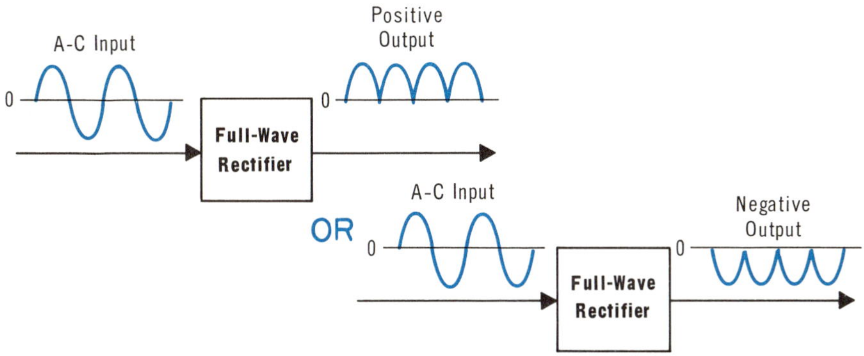

A rectifier can be designed to deliver either a positive or a negative output voltage

The output of a rectifier is developed across a resistive load, represented in the previous examples by load resistor R. The amplitude of the output voltage depends on the current flowing through the load resistance. The polarity, since it is a rectified voltage, is always the same. However, although the polarity is always the same, it can be either a positive or a negative voltage, depending on the desired output characteristics.

It should be understood here that this choice of output polarity is with respect to *ground*. The basic characteristics of a rectifier are such that it has a *negative side,* which in a full-wave rectifier is the transformer secondary centertap, and a *positive side,* which is the cathode side of the load resistance. This is shown in A on the following page. Either one of these sides can be connected to ground to provide a common reference potential.

If the *negative side* of the rectifier is grounded, the rectifier output is a *positive voltage*. This is shown in B on the following page. The voltage at the top of the load resistance increases and decreases, but is always at or above ground potential. If the cathode, or *positive,* side of the load is grounded, the output voltage is *negative,* as shown in C. This is because the positive side of the load remains at constant ground potential, so the varying voltage produced across the load is reflected as a changing potential at the negative side of the load. This changing negative potential with respect to ground is a negative voltage.

As shown in D, if the load resistance is split and the center point grounded, both a positive and a negative output can be obtained. Of course, the polarities can also be changed by reversing all of the diode connections.

output polarities (cont.)

A

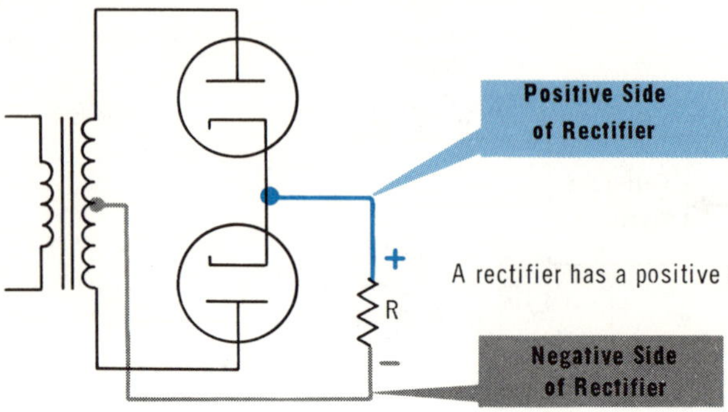

A rectifier has a positive and a negative side

B

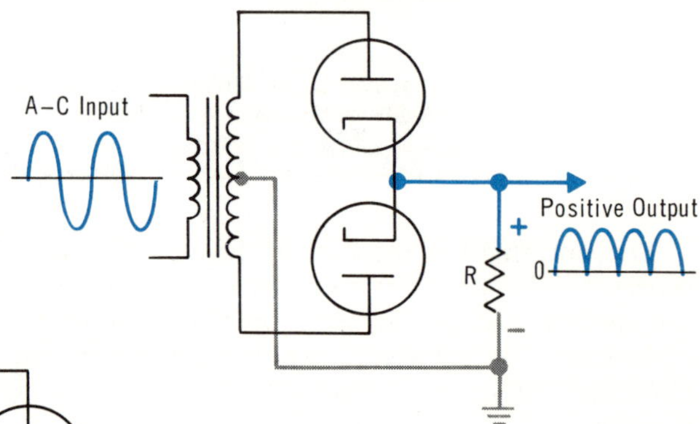

If the negative side is grounded, the output voltage is positive

C

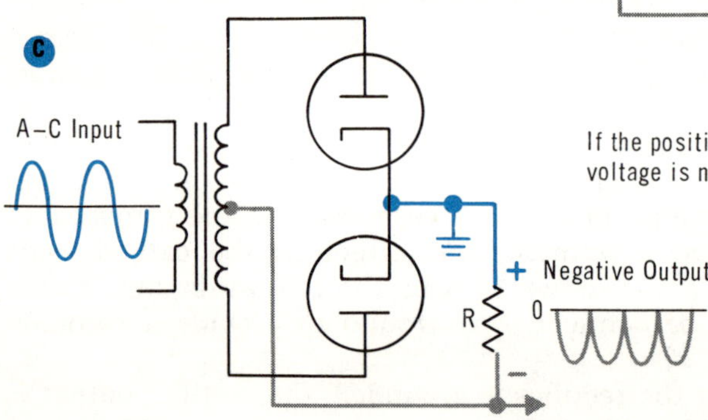

If the positive side is grounded, the output voltage is negative

D

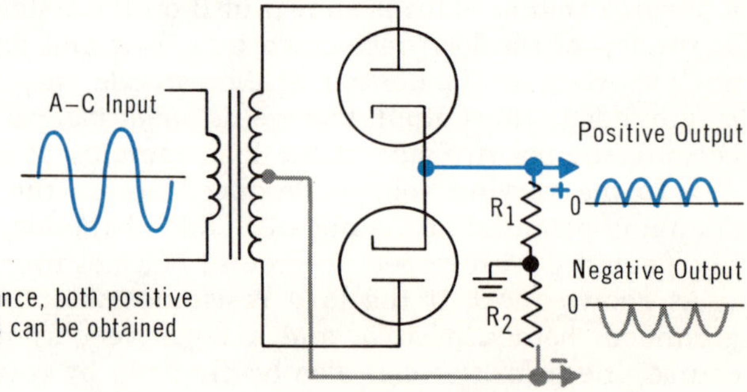

By using a split load resistance, both positive and negative output voltages can be obtained

summary

□ An electronic power supply converts a-c input power into one or more d-c voltages. It does this by first rectifying the sinusoidal ac, and then smoothing, or filtering, the resulting fluctuating dc. □ A half-wave rectifier delivers an output during only half of the input voltage cycle. □ A simple half-wave rectifier consists of a tube or semiconductor diode and a load resistor. The diode conducts only when its plate, or anode, is positive with respect to its cathode. □ Half-wave rectifiers are limited in the amount of output current that they can deliver. In addition, their output voltage fluctuates widely, requiring a considerable degree of filtering if a smooth dc is required.

□ The full-wave rectifier delivers an output during both halves of the input voltage cycle. □ A basic full-wave rectifier consists of two diodes having a common load. The diodes conduct during alternate half cycles of the input and both produce current through the load in the same direction. □ Since a full-wave rectifier produces an output pulse for every half cycle of the input, its output generally requires much less filtering than a half-wave rectifier.

□ A bridge rectifier is a special type of full-wave rectifier. It consists of four diodes and a common load. □ The output voltage produced by a bridge rectifier is nearly twice as large as that produced by a comparable conventional full-wave rectifier. □ The peak inverse voltage across each diode of a bridge rectifier is only half that which occurs across each diode of a conventional full-wave rectifier. □ In a rectifier circuit, the input a-c voltage is often increased by being applied through a step-up power transformer. □ The output of a rectifier circuit can be either a positive or a negative voltage, depending on which side of the rectifier is grounded.

review questions

1. Describe the operation of a vacuum-tube half-wave rectifier.
2. Draw a diagram of a full-wave semiconductor rectifier.
3. How would you change to output polarity?
4. What is meant by the peak inverse voltage of a rectifier?
5. Draw a diagram of a bridge rectifier.
6. Why are transformers used in power supplies?
7. What is the rms value of a full-wave rectified waveform?
8. What is the rms value of a half-wave rectified waveform?
9. State one advantage that a half-wave rectifier has over the bridge rectifier.
10. If the peak voltage across a transformer secondary is 100 volts, what is the peak value of a half-wave rectified voltage? Full-wave? Bridge?

filter circuits

You have seen on the previous pages how various types of rectifiers convert an a-c input into a fluctuating d-c output. The fluctuation of the d-c output above and below its average value is called *ripple*. In a *half-wave* rectifier, the frequency of the ripple is the *same* as the frequency of the a-c input, because it produces one pulse per cycle. In a *full-wave* rectifier, the ripple frequency is *twice* that of the a-c input, because two pulses per cycle are produced. Thus, if the input is from a 60-cps source, a half-wave rectifier has a ripple frequency of 60 cps, and a full-wave rectifier has a ripple frequency of 120 cps.

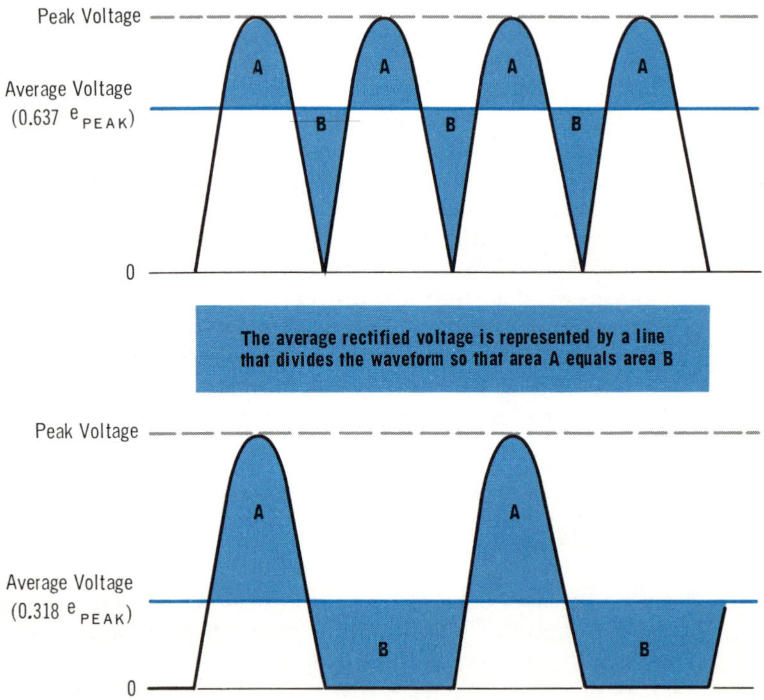

Most electronic equipment requires smooth d-c operating voltages. The output of a rectifier cannot be applied directly to such equipment because of the ripple. Consequently, the ripple must be eliminated. Circuits for accomplishing this are called *filter circuits*.

Essentially, a filter circuit compresses the peaks of the individual, fluctuating d-c pulses and fills in the valleys between pulses. The net result is an essentially smooth d-c voltage. The more elaborate the filter circuit, the more closely the output resembles the perfect dc of a battery.

the capacitance filter

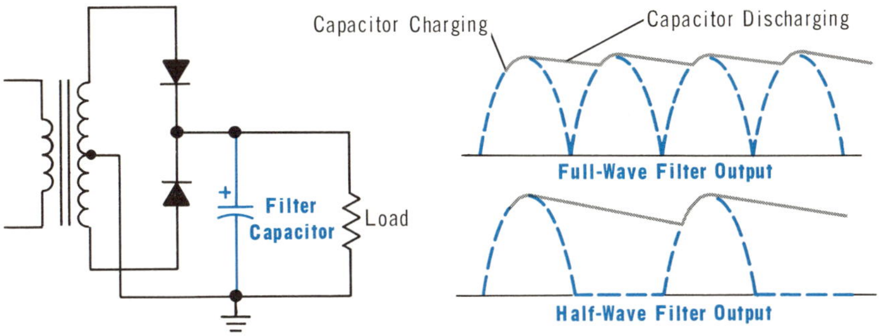

The discharge of the filter capacitor is greater between output pulses from a half-wave rectifier than it is from a full-wave rectifier. This means that the half-wave rectifier has poorer filtering regulation, since the filtered output voltage decreases between rectified pulses

The ripple output of a rectifier represents energy being supplied to the load in pulses. The ripple fluctuations can be reduced considerably if some of the output is *stored* while the rectifier is delivering a pulse and then *released* to the load between output pulses. This is the basic operating principle of the simple capacitance filter.

As shown, such a filter consists of a capacitor connected across the rectifier output. The capacitor charges rapidly, usually within a few cycles, to the *peak* rectifier voltage. When the rectifier output drops to zero between output pulses, the capacitor voltage supplies the load. The capacitor voltage also begins to fall off, since the capacitor starts to discharge through the load.

However, the rate of discharge is long compared to the time between rectifier output pulses. So, instead of falling off to zero between output pulses, the voltage applied to the load is maintained, although at a gradually diminishing level. Before the capacitor voltage drops too low, another output pulse is delivered by the rectifier. This recharges the capacitor to the peak voltage, so the capacitor is again ready to supply the load voltage when the rectifier output drops.

The rate at which the capacitor discharges, and, therefore, the average level at which the output voltage is maintained, depends on the RC time constant of the capacitor and the load resistance. For low values of load resistance, the discharge is fairly rapid, so the output voltage falls rather quickly between rectifier pulses. Such a condition, known as poor *regulation,* is highly undesirable. For this reason, the simple capacitor filter is not used with rectifiers that must supply a large load current, or, in other words, rectifiers that have low load resistances.

Mostly, *electrolytic* capacitors are used in filter circuits. These are special capacitors that are polarized. Usually, the positive side of the capacitor is marked so that it can be connected properly. If it is hooked up backwards, it will be destroyed, and might burst.

the inductance filter

An *inductor* is another device that can alternately store and release electrical energy. It does this by extracting energy from a flowing current and storing it in a magnetic field when the current is increasing. Then it releases the energy to keep the current flowing when the current begins to decrease. This ability of an inductor to store and release energy can be used to help prevent the abrupt changes in the output of a rectifier. To do this, the inductor is inserted in the rectifier circuit in *series* with the load, as shown. The inductor tends to prevent the output current from building up or dying down. Its effect on the output waveform is as shown.

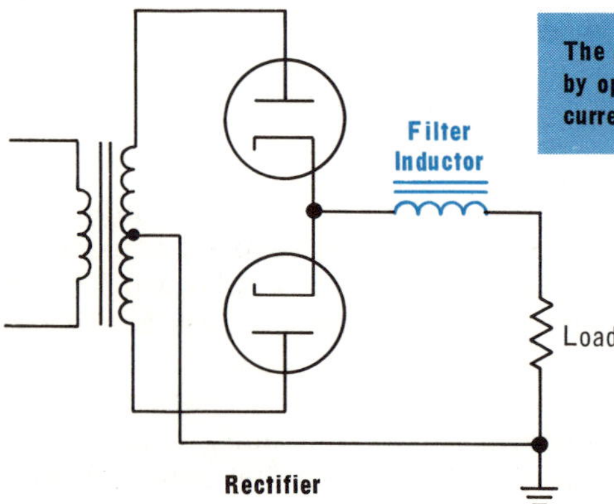

The inductance filter provides filtering action by opposing changes in the rectifier output current

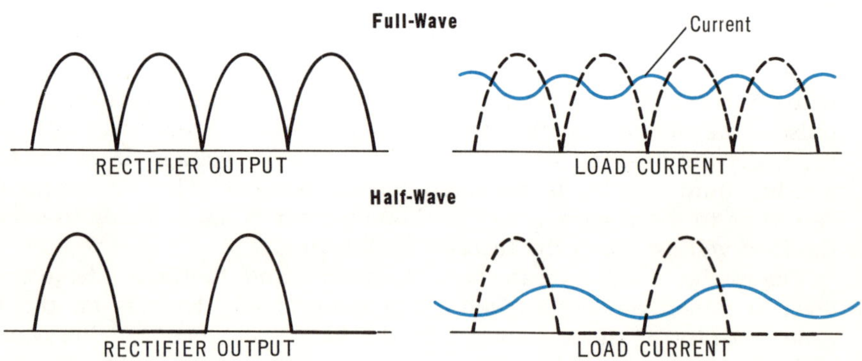

The inductance filter prevents the current, and, therefore, the output voltage, from ever reaching the peak value that would be obtained if the inductor were not in the circuit. Therefore, a rectifier that has an inductance filter will not produce as high an output voltage as one that has a capacitance filter. However, a larger load current can be drawn from the inductance filter without seriously changing the output voltage.

the capacitor-input filter

Neither the simple capacitance or inductance filter is capable of removing ripple voltage to a high degree. Consequently, when effective filtering is required, capacitance and inductance filtering are used together. One such *LC* filter is shown below. This is called the capacitor-input filter, because a capacitor (C_1) is the *first* filtering element directly after the rectifier.

The capacitor-input filter is probably the most widely used type of filter in applications involving low-current devices, such as radio receivers. It is also called the "pi" filter because of its schematic resemblance to the Greek letter π

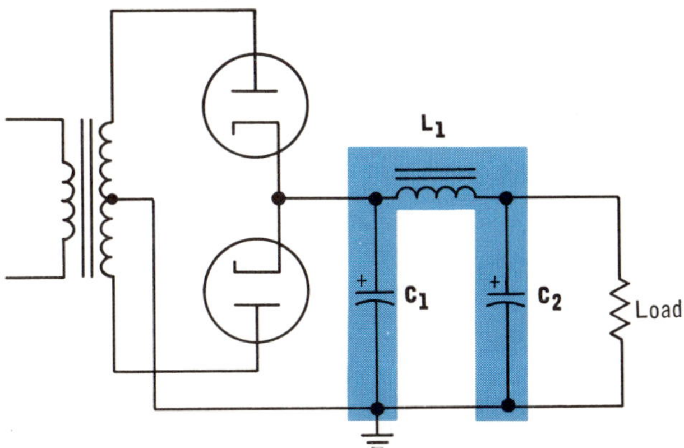

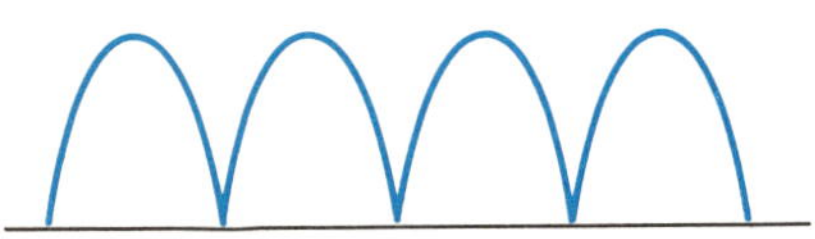

RECTIFIED VOLTAGE

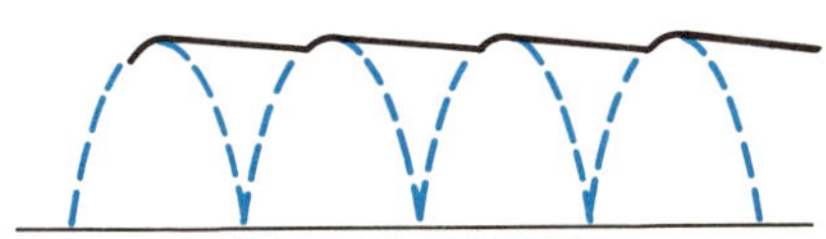

FILTERED VOLTAGE

In the capacitor-input filter, capacitor C_1 performs the same function as the simple capacitor filter previously described. It charges on the peaks of the rectified output pulses and then discharges, supplying energy to the load, when the rectifier output falls. Capacitor C_2 provides similar filtering action, although to a lesser degree. Inductor L_1 adds to the overall filtering action by opposing changes in both the output current filtered by C_2 and the current drawn by the load. The actions of L_1 and of C_2 are closely interrelated, and their values are chosen in conjunction with one another.

The output of a well-designed capacitor-input filter contains only a small amount of ripple, which for most practical applications can be ignored. However, the voltage regulation of such a filter is relatively poor. This is because of the decrease in the voltage across C_1 as it discharges between rectified pulses, as well as the d-c voltage drop produced by the load current flowing through L_1.

the choke-input filter

When an inductor is used as the primary filtering element in an LC filter, the network is called a choke-input filter. The term *choke* is used because of the inductor's ability to stop, or choke, the passage of ripple voltage to the load.

A simple choke-input filter is shown in A. The inductor opposes current changes, while the capacitor charges and discharges in standard filter-capacitor fashion. The combined effect of the two in reducing ripple is much greater than when either is used alone. However, since the inductor reduces the peak rectifier current, and this in turn reduces the maximum voltage to which the capacitor charges, the output voltage of the choke-input filter is *lower* than that of the capacitor-input filter. This output voltage, though, is affected less by changes in the load current. Consequently, choke-input filters are widely used in applications involving large, changing load currents.

When a higher degree of filtering is required, two choke-input filters can be connected in series. This is shown in B.

A

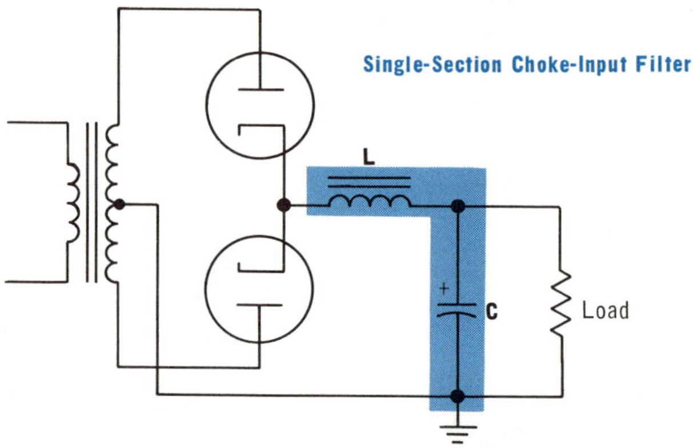

The output voltage from a choke-input is approximately the average value of the a-c voltage applied to the rectifier

B

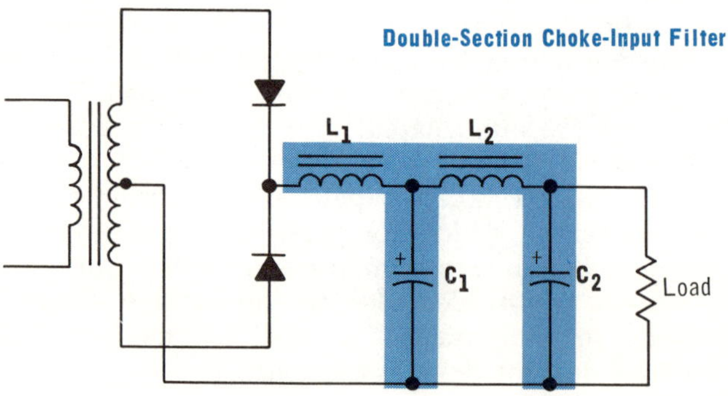

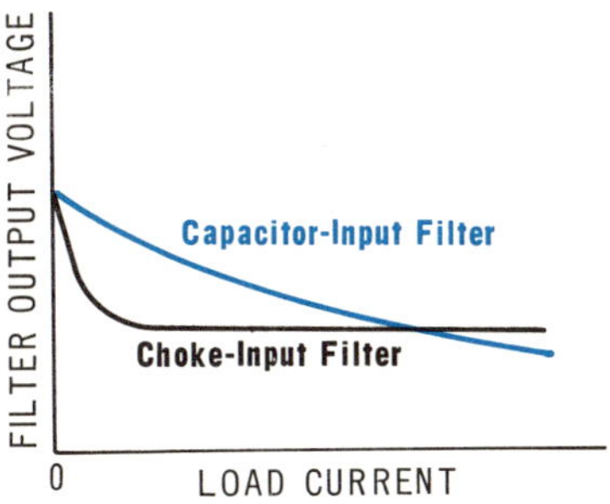

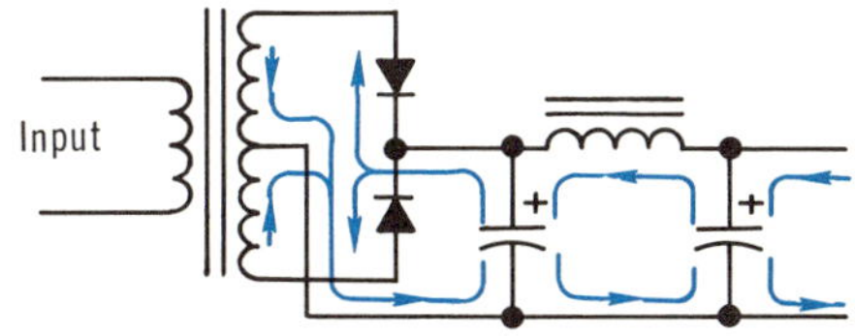

The output voltage from a choke-input filter is considerably more constant over a range of load current than is the output from a capacitor-input filter, because the charge on the capacitor is reduced as the load draws current

capacitor vs. choke-input filter

In a choke-input filter, current is supplied to the load through rectifier conduction directly. But with a capacitor-input filter, the current is supplied to the load primarily by the capacitor. The rectifier conducts to charge the capacitor. If the load did not draw current, and there was no bleeder resistor, the rectifier would not conduct because the voltage on the capacitor would put the cathode of the rectifier at the same potential as the anode. However, the load does draw current, which the capacitor discharges to supply, and the rectifier conducts to recharge the capacitor. Normally, the rectifier conducts only brief bursts of current to cause a pip-type waveform.

With very little current being drawn, the capacitor-input filter delivers an output voltage almost equal to the peak value of the a-c input to the rectifier. If the load draws an increasing amount of current, however, the output voltage falls because the rectifier cannot conduct long enough to recharge the input capacitor. To provide a sufficient charge on the capacitor would require a very large peak-rectifier current, which might damage the rectifier diodes. Consequently, capacitor-input filters are only used where *small* or *constant* load currents are involved.

With *no load* current being drawn, the output of the choke-input filter is also nearly equal to the peak value of the rectifier's a-c input because the capacitor has time to charge to peak. But, when a small amount of load current flows, the output voltage drops sharply to the average value of the a-c input because of the inductive action of the choke. As the load current is increased further, though, there is very little change in the output voltage. Thus, the choke-input filter is highly suitable for *heavy* or *variable* load current applications. Usually, a bleeder resistance is made a part of the choke-input filter. Thus, some load current always flows, preventing the large output voltage change that would otherwise occur between no-load and small-load conditions.

the swinging-choke input filter

The choke-input filter delivers an essentially constant voltage output over a wide range of load current. However, when the load current becomes *very heavy,* the output voltage starts to fall significantly. This is due to the d-c voltage drop the current produces across the inductor and power transformer. To counteract this decrease in voltage at heavy currents, a type of inductor called a *swinging choke* is often used.

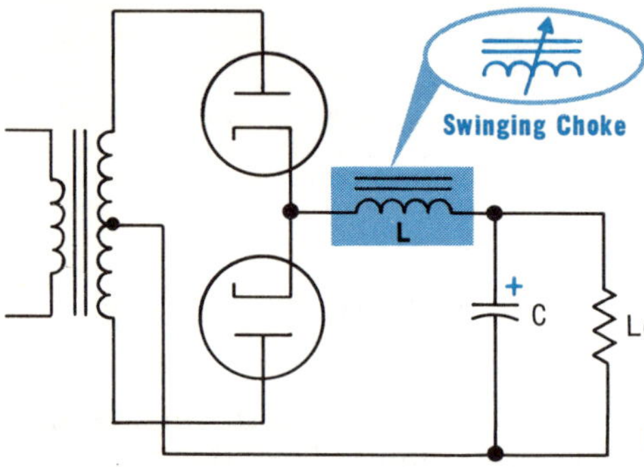

Up to a certain value of load current, the swinging choke acts as a conventional choke

Moderate Load Current

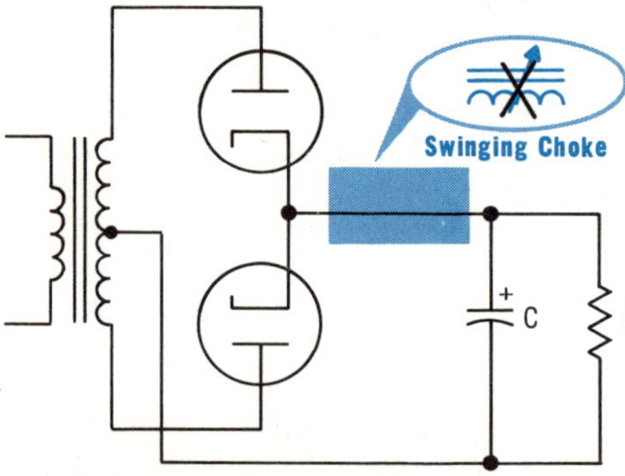

As the current increases further, the inductance of the choke falls until, at the maximum current level, the choke has no inductance, and the filter acts as a capacitor input

Heavy Load Current

Such an inductor does not have a constant inductance value. Instead, the inductance is relatively constant until a certain current flow is reached. Then, the inductance *decreases* as the current increases. This characteristic is due to the type of iron core used in the swinging choke. The choke actually starts saturating at a certain current level, and saturates more and more as the current goes up, so that the effective inductance continues to go down accordingly.

The swinging-choke input filter acts as a conventional choke-input filter below the point where saturation starts. Above the saturation point, the current increases, the inductance of the choke falls proportionately, and the filter becomes more and more like a capacitor-input filter. So, the capacitor can charge to a higher voltage, thus compensating for the decrease in output voltage that would otherwise occur because of the d-c losses in the choke and transformer.

the RC filter

In applications where less filtering can be tolerated, an RC filter is sometimes used. As shown, such a filter is similar to the capacitor-input filter, except that a *resistor* is used in place of the inductor. Although the resistor aids in the filtering by increasing the discharge time of capacitor C_1 and the charge time of capacitor C_2, it has a considerably higher *d-c resistance* than an inductor. Because of this, the RC filter is not suitable when large load currents must be supplied. Such large currents would produce an excessive voltage drop across the resistor, which would cause the output voltage to drop significantly.

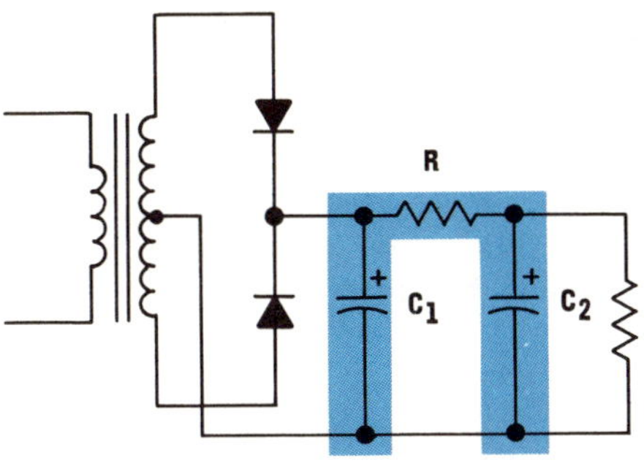

The output of the RC filter contains more ripple than when an inductor is used. This is because the filtering action of the resistor is not as effective as that of an inductor

RECTIFIED VOLTAGE FILTERED VOLTAGE

Because of the higher resistance of the resistor, the output of the RC filter drops a great deal as load current goes up

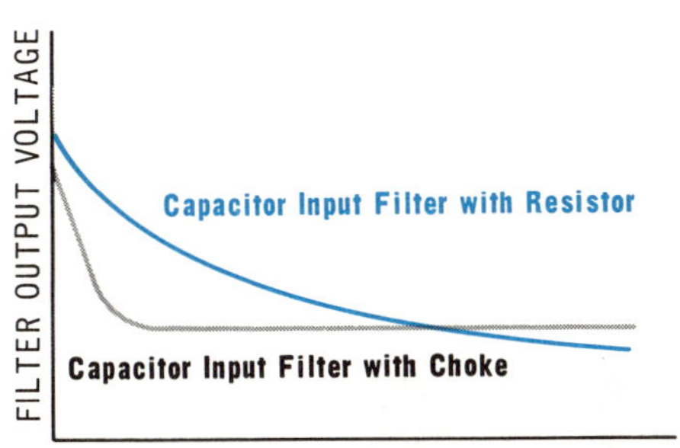

Although they are not used extensively in power supplies, RC filters are commonly employed as *decoupling networks* in amplifiers. The values of the RC components are different when they are used for decoupling purposes, because of the much higher frequencies to be filtered. However, the basic filtering principles are the same.

the effect of frequency on rectifiers and filters

Most electronic power supplies are designed to operate from a-c power sources that have a frequency of 60 cps. However, for frequency ranges of the a-c input that are greater than 60 cps, the rectifier and its filter can be made smaller and lighter. This characteristic is quite important in aircraft and other mobile applications, where size and weight are significant considerations.

Basically, the *reactive components* in the power supply are the elements responsible for this size and weight reduction at higher frequencies. When the frequency is increased, the reactance of an inductor increases, while that of a capacitor decreases. Consequently, at higher frequencies, smaller components are capable of developing the same filtering action obtained with larger components at lower frequencies. Similarly, power transformers having the necessary characteristics can be made both smaller and lighter as the frequency is increased.

Mobile power frequencies often run from 400 to 1000 cps.

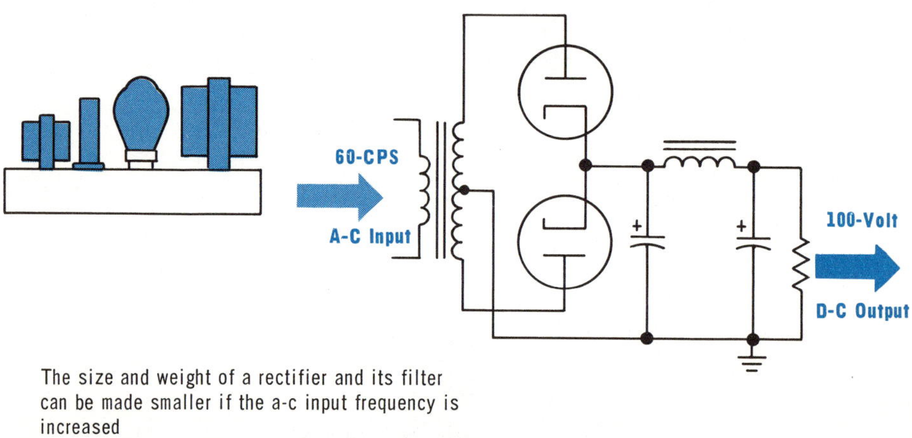

The size and weight of a rectifier and its filter can be made smaller if the a-c input frequency is increased

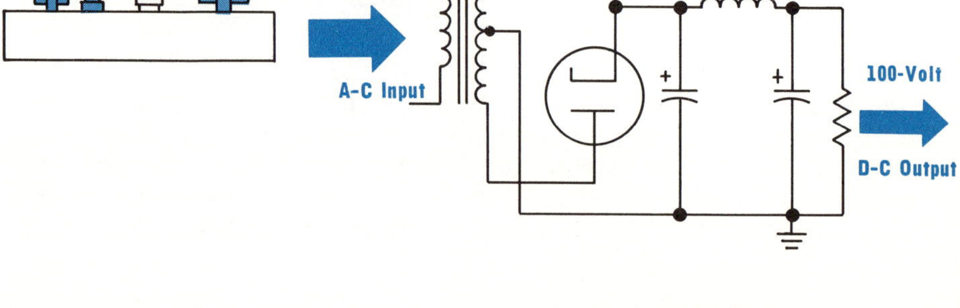

summary

□ Fluctuation of the d-c output from a rectifier above and below its average value is called ripple. □ The ripple frequency for a half-wave rectifier is the same as the input frequency. □ For a full-wave rectifier, the ripple frequency is twice the input frequency. □ A simple capacitance filter consists of a capacitor connected across the rectifier output. Such a filter is not used with rectifiers that must supply large load current, since it has poor regulation. □ The inductance filter consists of an inductor connected in series with the rectifier load. Although it will not produce as high an output voltage as the capacitance filter, it has better regulation.

□ The capacitor-input filter is a type of LC filter that can produce relatively large output voltages. Like all LC filters, it reduces ripple to a low value, but it has relatively poor voltage regulation. □ Choke-input filters, which are another type of LC filter, have good regulation. However their output voltages are less than those of capacitor input filters. □ Two choke-input filters are sometimes connected in series when a higher degree of filtering is required. □ The swinging-choke input filter uses a swinging choke, whose inductance decreases at heavy currents. This improves regulation when large load currents are drawn. □ The RC filter is similar to the capacitor-input filter, except that a resistor is used in place of the inductor. RC filters are used only for small current loads.

□ Power supplies operating at higher frequencies require smaller components for filtering action. This is because the reactances of the components vary with frequency. □ Mobile power frequencies run from 400 to 1000 cps.

review questions

1. Draw a diagram of a capacitor-input filter. Explain its operation.
2. What is *ripple*?
3. Is a ripple voltage a d-c or an a-c voltage? Explain.
4. Why does an RC filter have poor voltage regulation?
5. The input frequency to a bridge rectifier is 500 cps. What is the ripple frequency?
6. Would the ripple frequency change if the bridge rectifier had a swinging-choke-input filter connected to it?
7. Which has a greater ripple frequency: a half-wave rectifier whose input is 500 cps, or a full-wave rectifier whose input is 300 cps?
8. Why are electrolytic capacitors used in filters?
9. Why is electrolytic *polarity* important?
10. Why are high input frequencies used for mobile equipment?

the load resistor

In all the schematic illustrations shown previously, a resistor has been connected across the output of the filter. This resistor has represented the total resistance of the load being supplied by the power supply. In practice, an *actual* resistor is almost always placed across the filter output. Sometimes, the purpose of this resistor is to provide a fixed load for the power supply even when the external circuits being supplied draw no current. Such a resistor, called a *load resistor,* has a resistance value such that it draws about 10 percent of the full-load current. The resistor helps regulate the voltage when little load current is drawn.

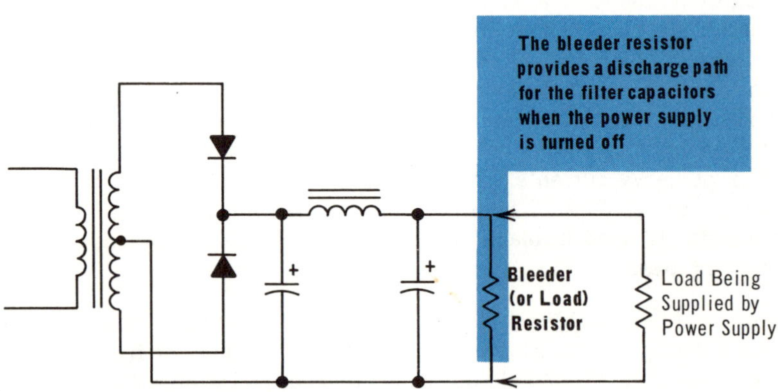

Without the bleeder resistor path, dangerously high voltages could exist at various points in the equipment for some time after it is deenergized. The bleeder resistor sometimes also provides a fixed load for the supply to improve regulation

As mentioned earlier, in a choke-input filter the output voltage drops sharply from no load to a small load. When a load resistor is used, there cannot be a no-load condition, since the resistor is always drawing current. Thus, there will be no initial sharp drop in the curve. During operation, the resistor helps regulation because it is supplying a current that causes an internal voltage drop in the supply. If the output voltage tends to go down for some reason, the current drawn by the resistor also goes down, causing somewhat less of a drop in the supply. This offsets some of the output voltage drop. This is explained more fully later.

If the resistor is used to provide a discharge path for the filter capacitors when the power supply is turned off, it is also called a *bleeder resistor*. When used only for this purpose, the resistor has a very high resistance value so that it will draw very little current.

Very often, the resistor serves as both a bleeder and a fixed load. Its resistance value is then selected accordingly.

the voltage divider

In many equipments, a single power supply provides two or more output voltages. This is done by using series resistors or a tapped resistor in place of a single load resistor. The rectifier current flowing through the series resistors produces a total voltage drop equal to the filter output voltage. But this voltage is *divided* into separate voltage drops across the individual resistors. This is shown in A.

Resistors R_1, R_2, and R_3 form a voltage divider across the filter output. With no external load connected and 300 volts at the output of the filter, 2 milliamperes of current flow through the voltage divider network. This is found simply with an application of Ohm's Law. One hundred volts is therefore dropped across each resistor.

Normally, some point on the voltage divider is grounded, and all voltages are measured with respect to that point. Various output voltages can then be obtained, depending on what point is grounded. This is shown in B and C.

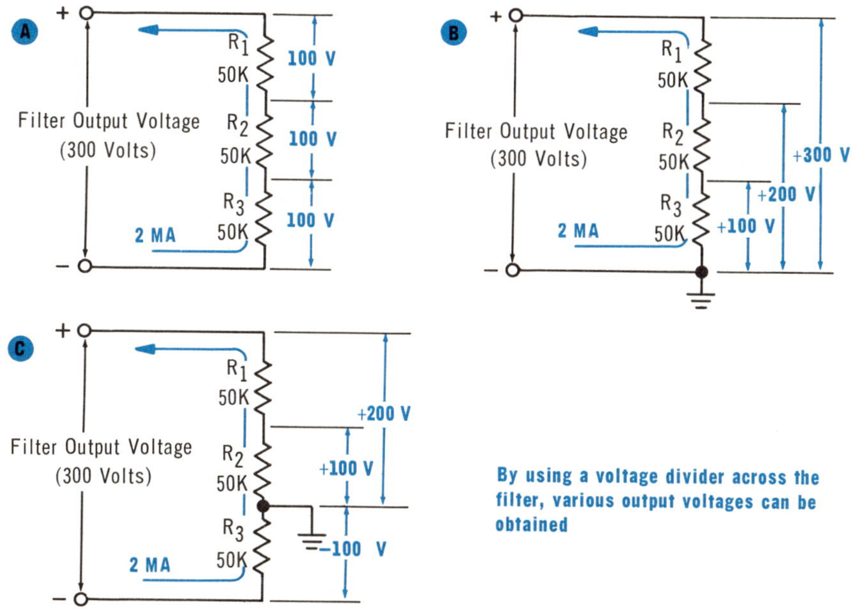

By using a voltage divider across the filter, various output voltages can be obtained

In A, B, and C, it is assumed that *no external load* is connected to the voltage divider. If a load, or loads, is connected, the voltage divisions shown are not correct. This is because the resistance of the load forms a parallel circuit with the portion of the divider it is connected across. This changes the resistance of that portion of the divider, and, therefore, affects the total resistance, the current, and all of the voltages. For this reason, such voltage dividers are designed for the particular load conditions under which they will operate.

voltage multipliers

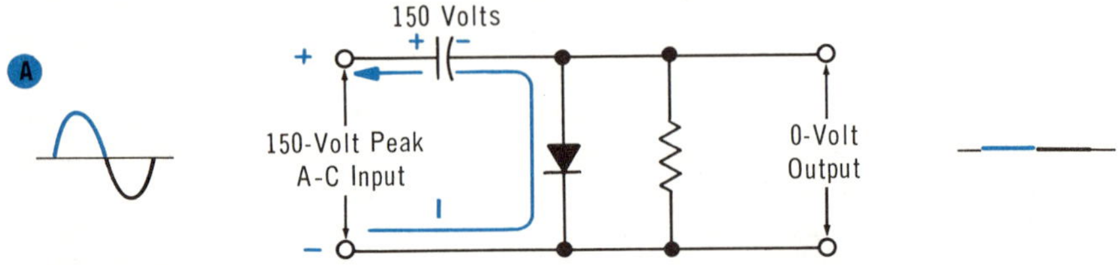

In this simple half-wave voltage doubler, the capacitor and input voltages alternately add and cancel

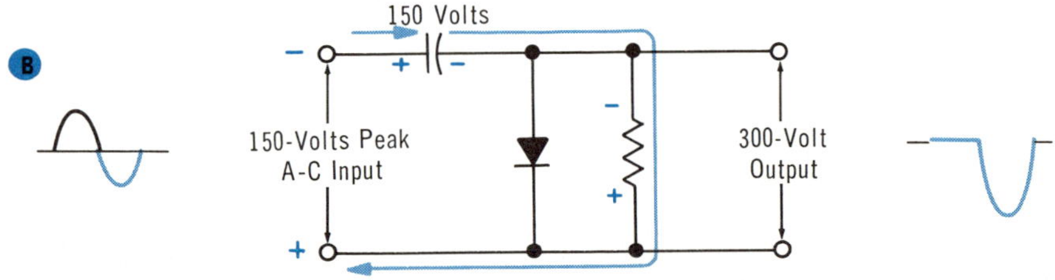

A voltage multiplier is a circuit that delivers an output voltage that is some *multiple* of the peak input voltage. For example, a *voltage doubler* has an output voltage that is *twice* the peak value of the input voltage.

The simplest type of voltage doubler is shown. It consists of a diode and a capacitor connected in series across an a-c input source. During the positive half cycle of the a-c input (shown in A), the diode conducts, charging the capacitor with the polarity shown. The output of the circuit is taken from across the capacitor *in series* with the input terminals. Since the input voltage and the capacitor voltage have equal amplitudes, but opposite polarities, the output is zero during charge time.

During the negative half of the input cycle (shown in B), the capacitor retains its charge, since the diode will not conduct in the reverse direction. The capacitor voltage and input voltage are now of the same polarity, so they add to produce a peak output voltage of 300 volts. The capacitor, therefore, doubles the applied voltage during this half cycle. This is why it is often referred to as the *doubling capacitor*. The output voltage, then, varies between zero on one half cycle, to twice the peak of the a-c input during the other half cycle. Such a circuit is called a *half-wave doubler*.

The simple voltage doubler shown here is not very practical. Not only does the output drop to zero during the positive half cycles of the input, but any current drawn by the load during the negative half cycles discharges the doubling capacitor, thus reducing the output voltage considerably.

a practical half-wave doubler

In order to make the previous circuit more practical, the output must be prevented from dropping to zero when the charge on the doubling capacitor opposes the input voltage, and the load must be kept from draining the doubling capacitor. This is done: (1) by adding another capacitor to store the output charge and supply current to the load, and (2) by adding another diode to isolate the input circuit when the doubling voltage cancels the input voltage. This is shown in the circuit on this page.

Capacitor C_1 and diode CR_1 function the same as the simple doubler circuit just described. Capacitor C_2 was added to charge to the output voltage, and diode CR_2 was added to isolate C_2 when the charge on C_1 cancels the input voltage. During the half cycle when the charge on C_1 *aids* the input voltage, the polarity of CR_2 is such that it is biased in the *forward* direction, and so allows C_2 to charge to the doubled voltage. On the other half cycle, when the charge on C_1 *opposes* the input voltage, diode CR_2 is *reverse biased,* and so offers a high resistance. It effectively opens the circuit to C_2, so that even though the sum of the charge on C_1 in series with the input voltage equals zero, capacitor C_2 still applies the doubled voltage to the load. Although C_2 does discharge somewhat to supply current to the load, its charge is replenished during the next half cycle.

Since C_2 can only charge during alternate half cycles, high current drains will cause C_2 to discharge considerably to lower the output voltage. Thus, this is a half-wave doubler circuit that is fit only for supplying a small output current. This circuit is also known as a *cascade* doubler, as will be explained later.

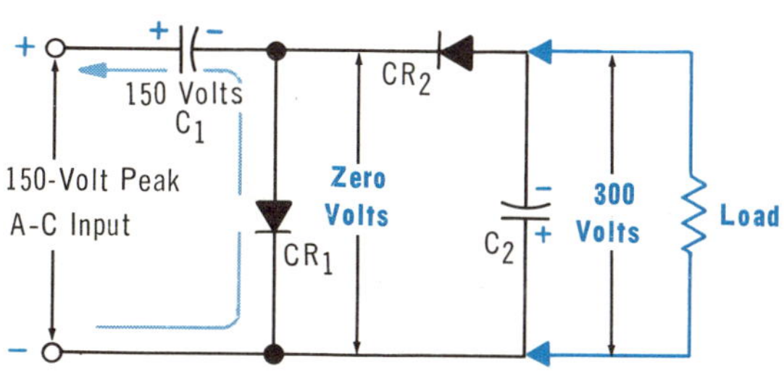

When C_1 is charging and its voltage cancels the input voltage, the sum of zero volts applied to CR_2 reverse biases CR_2, which opens the circuit to C_2. C_2 can then still supply its charge to the load

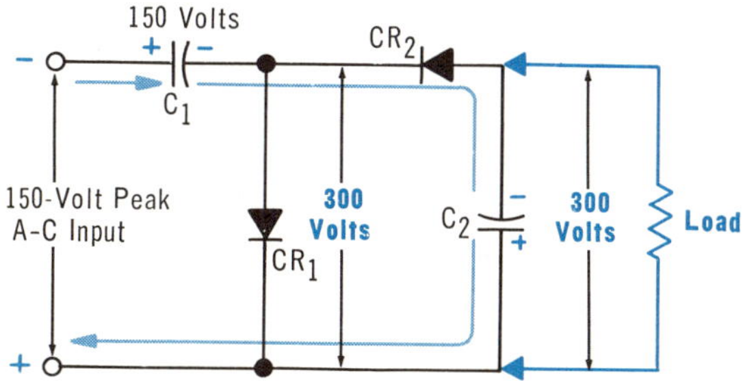

When the input polarity aids the charge on C_1, the doubled voltage is applied to CR_2. Then, when the charge on C_2 drops as it supplies current to the load, CR_2 conducts to recharge C_2

the full-wave doubler

A full-wave doubler circuit utilizes *both* cycles of the a-c input to replenish the output voltage. Essentially, two capacitors are charged on alternate cycles of the input and arranged so that their voltages *add* in the output. Such a doubler circuit is shown in A.

When point A on the transformer secondary is positive with respect to point B, diode V_1 conducts and charges capacitor C_1 to the peak value of the input, with the polarity shown. When the input reverses polarity and point B on the transformer secondary becomes positive with respect to point A, diode V_2 conducts. This charges capacitor C_2 to the peak value of the input, with the polarity shown.

Capacitors C_1 and C_2 are in *series* across the output. Their polarities are such that their voltages add. Since one capacitor is always being charged in any given half cycle, the voltage does not drop as much as it would for the half-wave doubler when the load draws current. Thus, this type of doubler circuit can be used for average load currents, but is not fit for heavy currents.

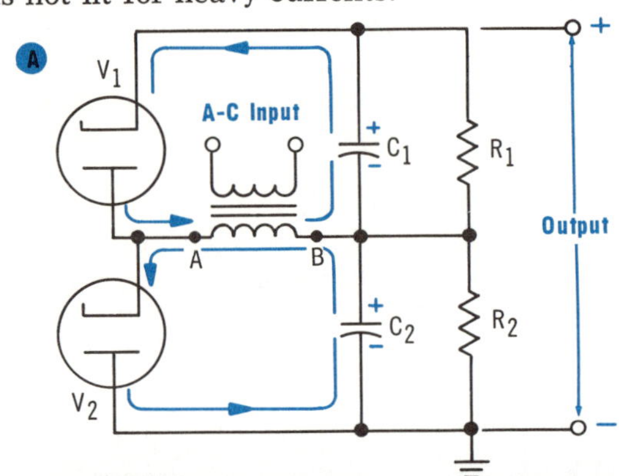

The full-wave rectifier draws current from the a-c source during both halves of the input cycle

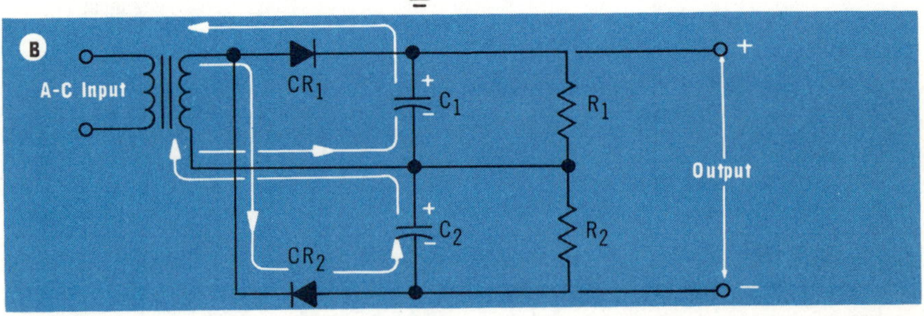

The semiconductor version shown in B is the same circuit, arranged differently to show how different schematic representations can be used. Diodes CR_1 and CR_2 conduct on alternate half cycles of the input, charging their respective capacitors, C_1 and C_2. The load is connected so that the two capacitors are in series across it. The output voltage is, therefore, twice the peak value of the input.

the cascade doubler

As you have learned, the basic half-wave doubler circuit is called a *cascade doubler*. The tube version of this circuit is shown connected for a positive output. When point A on the transformer secondary is negative with respect to point B, diode V_1 conducts. The conduction path is through capacitor C_1, which causes C_1 to charge to the peak value of the applied ac. When the polarity of the a-c input reverses, point A on the transformer becomes positive with respect to point B. The input voltage and the voltage across capacitor C_1 are now effectively in series between the plate and cathode of diode V_2. Therefore, V_2 conducts, and, in doing so, charges capacitor C_2 to *twice* the peak value of the a-c input.

The cascade doubler, as well as any other voltage doubler, can be used with or without an input power transformer

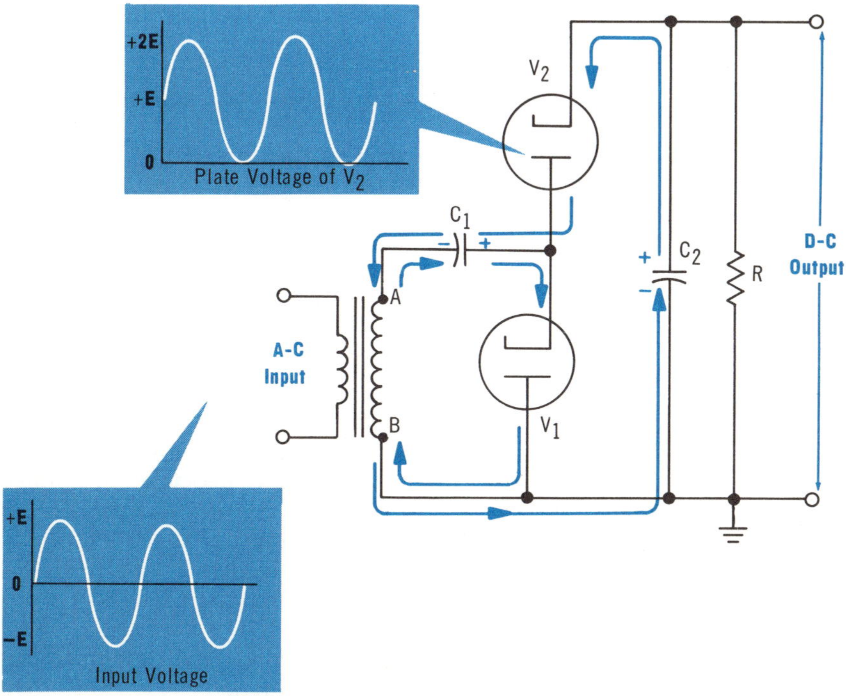

In effect, therefore, capacitor C_1 and diode V_1 cause the plate voltage of diode V_2 to vary between zero and twice the peak value of the input. This voltage is used to recharge capacitor C_2 when diode V_2 conducts. Resistor R serves as a bleeder or load resistor.

triplers, quadruplers, and quintuplers

Actually, voltage multipliers can be combined in various ways so that the output voltage is any multiple of the input. Many different methods can be used to do this. One simple method of making a tripler is shown on this page. A doubler and a simple rectifier are combined so that their outputs add to get three times the input voltage. Two doublers with the right polarities are combined to form a quadrupler.

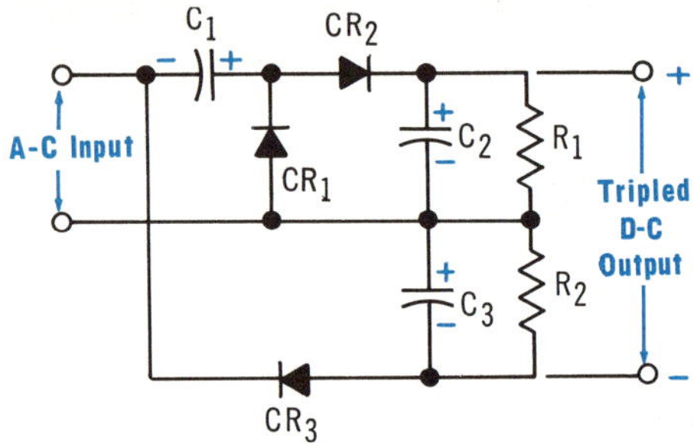

This voltage tripler consists of a cascade doubler and a half-wave rectifier, with their outputs (C_2 and C_3) in series

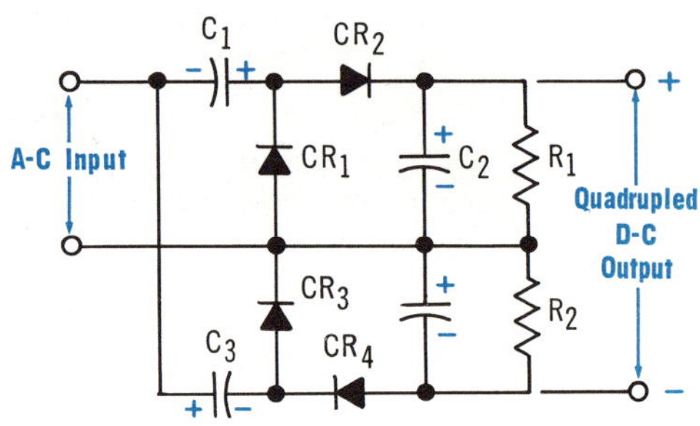

The quadrupler uses two doublers with their outputs in series

There is a type of cascade circuit that can be repeated to get any desired multiple of output. This is shown on the following page. Part A shows the basic rectifier. On one half cycle, V_1 conducts to charge C_1 to 100 volts. If the circuit is extended as shown in B, when the input polarity reverses, it adds to the charge on C_1, and 200 volts is applied to V_2, which conducts to charge C_2 to 200 volts. If you take the output across C_2, you have a doubler circuit. If the circuit is extended again, as shown in C, when the input polarity reverses again, it adds with the charge on C_2 to apply 300 volts to V_3, which charges C_3 to 300 volts. The output across C_3, then, gives us a tripler.

You can see in C that the charge on C_1 is also replenished at this time. Part D shows how a quadrupler section is added. When the input voltage is in series aiding with the charge on C_3, V_4 conducts and charges C_4 to 400 volts. With the quintupler section (E), when the input and the charge on C_4 add, V_5 conducts to charge C_5 to 500 volts.

This pattern can be repeated for any voltage multiplication desired, but it is generally inefficient after quadrupling. The voltage is not quite doubled each time because of losses.

triplers, quadruplers, and quintuplers (cont.)

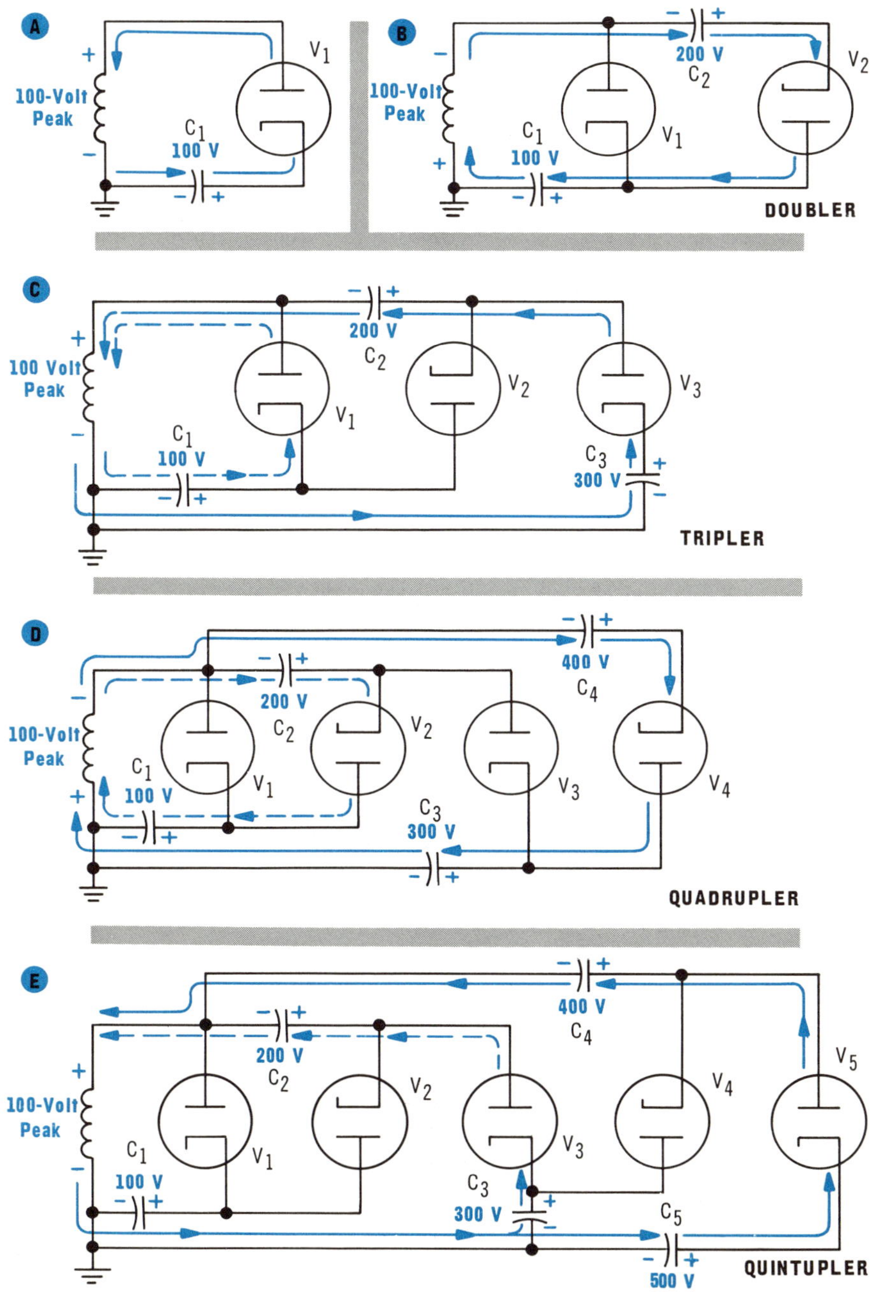

summary

☐ A load or bleeder resistor is usually connected across the output of a filter. ☐ The purpose of a load resistor is to provide a fixed load even when the external circuits do not draw current. This improves the supply regulation. ☐ A bleeder resistor is used to provide a discharge path for the filter capacitors when the power supply is turned off. ☐ Sometimes the load resistor is also used as a bleeder.

☐ A voltage divider is used to obtain more than one voltage from a single power supply. Usually a divider is a group of series resistors or one tapped resistor used in place of a load resistor. ☐ Normally, some point on the voltage divider is grounded, and all voltages are measured with respect to that point.

☐ A voltage multiplier is a circuit that delivers an output voltage that is some multiple of the peak input voltage. ☐ A simple half-wave voltage doubler consists of a capacitor and diode connected in series across the input ac. Such a doubler is not very practical, and must be modified to improve its regulation. ☐ The modified half-wave voltage doubler contains a second capacitor and diode, and is also known as a cascade doubler. ☐ The full-wave doubler utilizes both cycles of the a-c input to produce the doubled output voltage. ☐ Voltage multipliers can be combined in various ways so that triplers, quadruplers, and even quintuplers are obtained. However, any multiplier above a quadrupler is generally inefficient because of losses. ☐ The higher the multiplier, the poorer is the regulation of the system; even a small value of load current will cause a sharp drop in output voltage.

review questions

1. What is the purpose of a bleeder resistor?
2. What is the purpose of a fixed load resistor?
3. Can a fixed load resistor ever be used as a bleeder?
4. What is the purpose of a voltage divider?
5. Can a voltage divider ever yield a voltage greater than the input voltage?
6. What is a *voltage multiplier*?
7. Is there a limit to the voltage increase that can be obtained from voltage multipliers? Why?
8. Draw a diagram of a practical half-wave doubler. Explain its operation.
9. Draw a diagram of a full-wave doubler.
10. How can more than one polarity of voltage be obtained from a voltage divider?

vibrator power supplies

The vibrator power supply is a special kind of supply used to produce a high value of dc from a low-voltage d-c source. Usually, the low-voltage source is a storage battery. Vibrator supplies are used in such mobile equipment as the automobile radio.

The basic principle on which the vibrator supply works is illustrated here. A vibrating armature, or reed, is connected in series with a low-voltage d-c source and the centertap of a transformer primary. The armature is made to vibrate rapidly, alternately making contact with terminals 1 and 2. When the armature contacts terminal 1, current flows through the upper half of the transformer primary. When the armature contacts terminal 2, current flows through the lower half. The alternate currents in the primary halves induce voltages of opposite polarity in the transformer secondary. The secondary voltage is therefore continuous ac, having a modified square-wave waveshape.

Since a step-up transformer is used, the output developed across the secondary is considerably greater than the input voltage. The a-c output voltage is then rectified and filtered to produce the required d-c voltage. The manner in which the armature is made to vibrate is shown on the following pages.

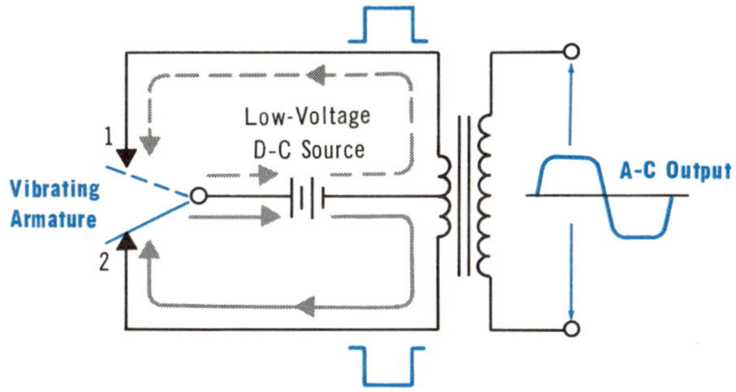

As the armature vibrates, current alternately flows in opposite directions in the upper and lower halves of the transformer primary. This induces a stepped-up a-c voltage in the secondary

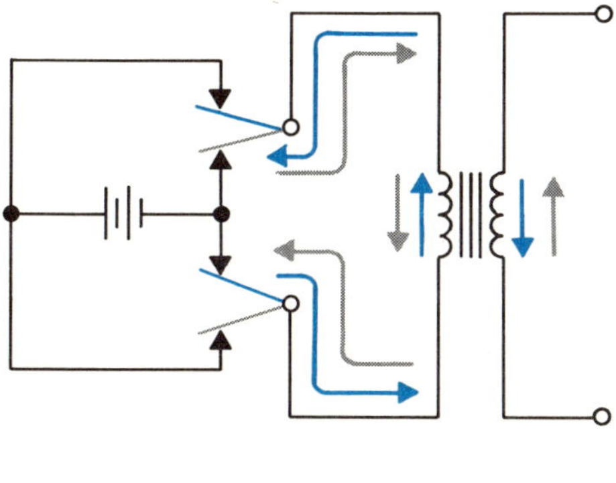

The same thing can be accomplished without a centertapped transformer. Double-pole, double-throw contacts switch the battery connections across the entire primary

the nonsynchronous vibrator power supply

A typical *nonsynchronous* vibrator power supply is shown. This is called a nonsynchronous supply because it requires a *rectifier* to produce the output dc. The vibrating armature is caused to vibrate between contacts 1 and 2 by the action of an electromagnetic relay, K. The armature is pulled against contact 1 by a spring when the vibrator is deenergized.

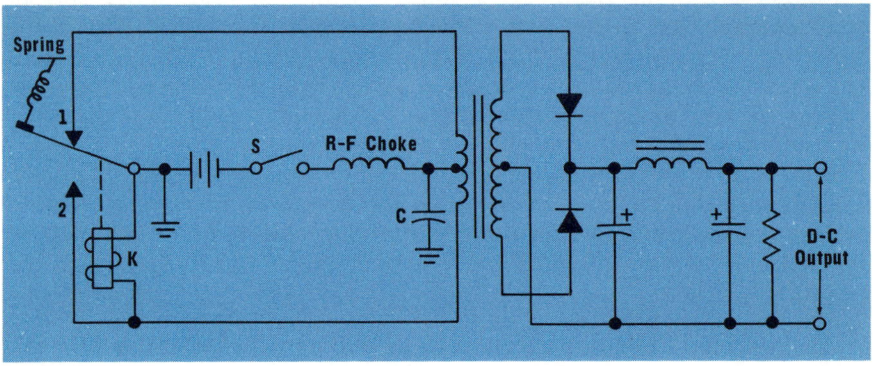

A nonsynchronous vibrator power supply requires a rectifier in the transformer secondary circuit

When vibrator switch S is closed, current flows from the battery through the r-f choke, the lower half of the transformer primary, and the relay, and back to the battery. The current energizes the relay, which pulls the armature to contact 2.

When the armature touches contact 2, the relay is shorted out of the circuit. As a result, it no longer attracts the armature, which is then returned to contact 1 by the spring. When the armature touches contact 1, current flows from the battery through the upper half of the transformer primary. This current is also of short duration, though, since when the armature leaves contact 2, current again begins to flow through the relay. When this current builds up sufficiently, the relay again pulls the armature to contact 2.

In this way, the armature vibrates rapidly between contacts 1 and 2. The resulting currents in the two halves of the transformer primary produce a stepped-up a-c voltage across the secondary. This voltage is rectified and filtered by a conventional full-wave rectifier and capacitor input filter.

The r-f choke and capacitor C form a *hash filter*. They prevent the electrical noise, or hash, produced by the sparking armature contacts from being passed to the power-supply output.

the synchronous vibrator power supply

A synchronous vibrator supply needs only a filter, since the vibrator and transformer produce a fluctuating d-c voltage

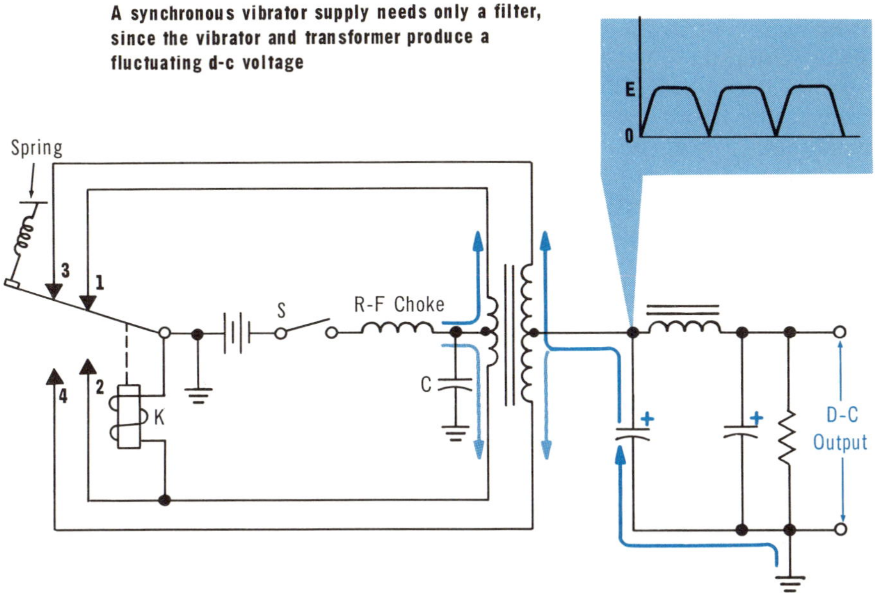

With every change in current direction in the primary, the secondary winding connections are reversed so that dc is applied to the filter

A *synchronous* vibrator power supply is one that does not require a separate rectifier to produce the d-c output. Such a vibrator supply delivers a *fluctuating d-c* output across the secondary of the transformer. All that is required is a filter network to smooth out the fluctuations in the output.

A typical synchronous vibrator supply is shown. You can see that it is similar to the nonsynchronous supply, except that it has two additional vibrator contacts (3 and 4), each connected to the *opposite* side of the transformer secondary. When the armature touches contact 1, it also touches contact 3, when it touches contact 2, it also touches contact 4.

The extra contacts *ground* the end of the transformer secondary to which they are connected when their respective main contacts (1 or 2) are made. This causes all of the current pulses induced in the secondary to flow in the same direction. Thus, instead of an a-c secondary voltage, a fluctuating dc is produced. A conventional filter is then used to smooth the fluctuating dc. Essentially, every time the direction of current changes in the primary winding, which tends to alternate the voltage induced in the secondary winding, the connections to the secondary windings are reversed so that dc is applied to the filter.

multiphase (polyphase) power supplies

All of the electronic power supplies described thus far have operated from single-phase a-c input power. Power supplies that must deliver great amounts of d-c power, such as a kilowatt or more, generally operate from multiphase power sources. From your studies of electricity, you recall that such sources supply a-c power on three or more conductors, with the voltage between any two of the conductors equal but having *different phases*. In a multiphase, or *polyphase*, power supply, the voltages between the various conductors of the a-c input are rectified separately and developed across a *common output*. The output voltage is relatively simple to filter because the ripple frequency is higher than that of a single-phase rectifier.

Actually, for a half-wave rectifier, the ripple frequency equals the number of phases in the input times the fundamental input frequency. For example, if the input is 3-phase, 60-cps, the ripple frequency is 3×60, or 180 cps. For a full-wave rectifier, the ripple frequency would be twice this.

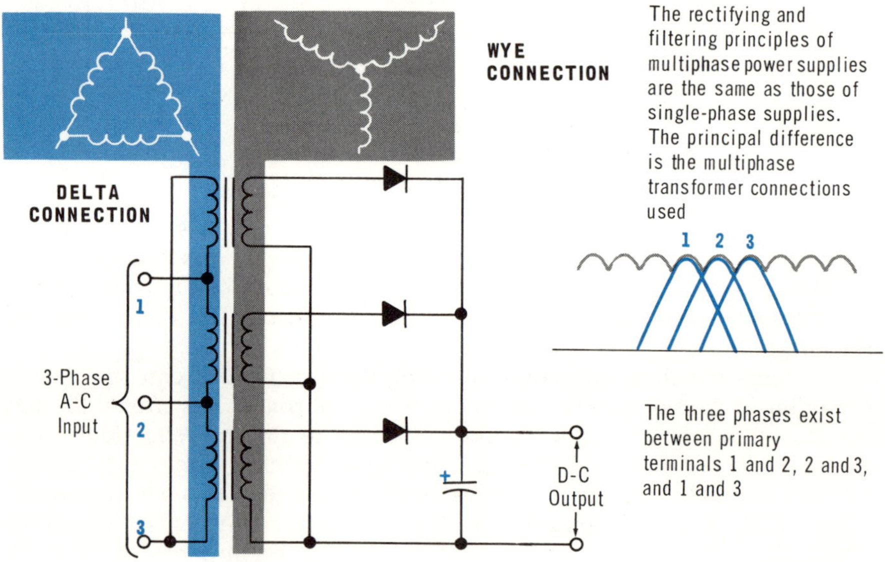

The rectifying and filtering principles of multiphase power supplies are the same as those of single-phase supplies. The principal difference is the multiphase transformer connections used

The three phases exist between primary terminals 1 and 2, 2 and 3, and 1 and 3

A basic 3-phase, half-wave power supply is shown. The primary windings of the power transformer are connected in the *delta* configuration, while the *wye* connection is used for the secondaries. Actually, a study of multiphase power supplies is in large part a study of multiphase transformer connections.

voltage regulation

You have seen how one very important characteristic of any power supply is the degree to which its output voltage remains *constant* despite changes in the amount of current drawn from the supply by the load. This characteristic is called the *regulation* of the supply. A supply whose output voltage changes little over a wide variation in load current is said to have *good* regulation. One whose output voltage falls sharply as the load current increases has *poor* regulation. Certain types of power supplies have inherently good regulation, mainly because of the type of filter used. However, for many applications, even these supplies do not have good enough regulation. Additional circuits must be added so that the output voltage remains constant. These additional circuits are called *voltage regulator* circuits.

Whenever the line voltage goes up or down, the power supply output tends to rise and fall, correspondingly. Also, a rising load current tends to reduce the output voltage, while a lower load current tends to cause the supply voltage to rise

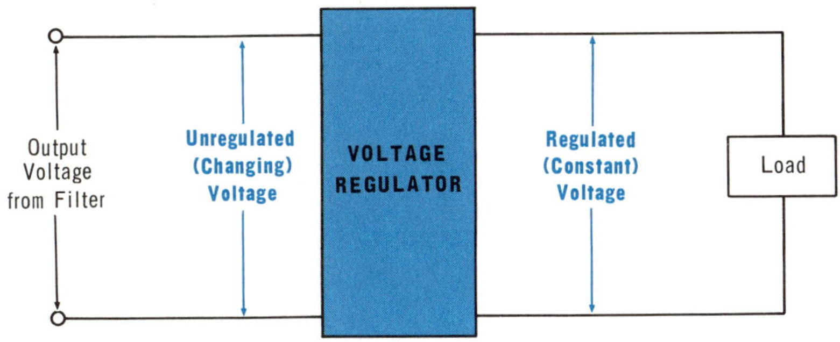

A voltage regulator is connected in the output of a power supply to maintain the voltage delivered to the load constant

Basically, the output voltage tends to vary because of two things: (1) fluctuating line voltage and (2) fluctuating load current. Every time the line voltage goes up, the output voltage will go up. But when the load current goes up, the output voltage will go down. This is because the output voltage is actually what is left after the internal voltage drops of the supply are taken away from the output. As the load increases, the internal drops increase and less is left for the output.

Generally, supplies are rated with percentages of regulation within certain load and line variations. For example, a 1% rating means that the output voltage will not vary more than 1 percent for load and line changes within specified limits.

principles

Voltage regulation is the last step in a power supply's conversion of a-c input power into an almost perfectly constant d-c output voltage. Increases in output voltage, which would otherwise be caused by a drop in the load current drawn from the supply or a rise in the level of the a-c input to the supply, are prevented from occurring by a voltage regulator. Decreases in output voltage, such as those caused by an increase in the load current drawn from the supply or a fall in the a-c input level, are also prevented.

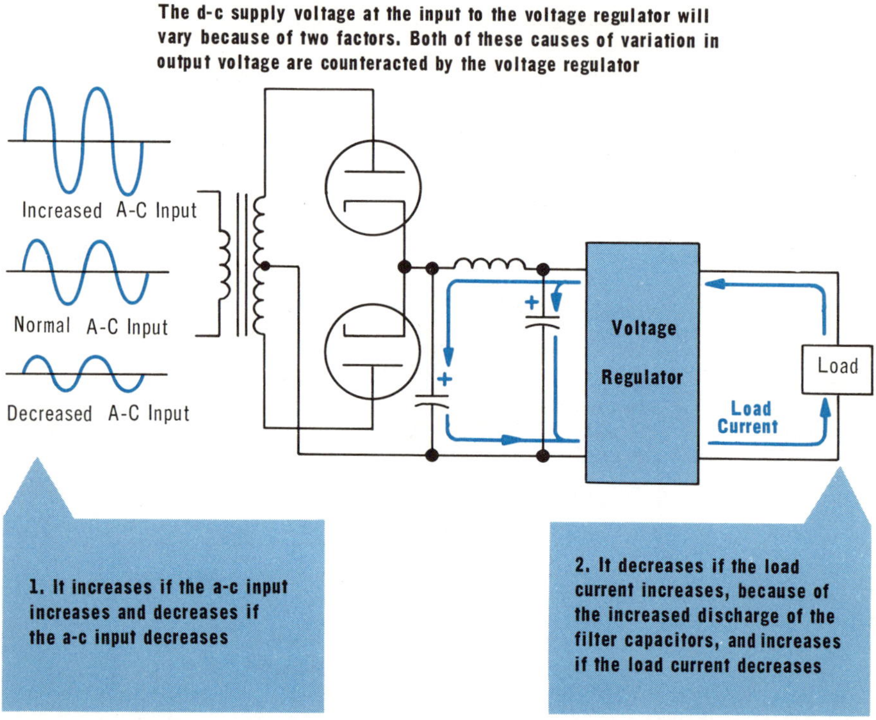

The basic voltage regulator circuit works by inserting resistance in the output of the supply, so that the output voltage is somewhat *less* than it would be without the regulator. Then, when the supply voltage varies, the resistance added by the regulator automatically changes to compensate for the voltage changes. In effect, the regulator circuit absorbs the voltage changes, dropping less of the output voltage when the voltage decreases and more when the output voltage increases. The net effect is that, after the voltage regulator, the supply output voltage is constant.

load resistor regulation

As was pointed out previously, when a load resistor is connected across the power-supply filter, it provides a small degree of regulation, especially when the load current is relatively small. Although the load resistor is an inefficient regulator, a description of the way it provides regulation is useful as an introduction to voltage regulators in general.

As shown, the load resistor is in parallel with the load; both are connected directly across the power-supply output. Under normal conditions, current flow through the supply is what is drawn by both resistor R and the load. Therefore, the current drawn by the resistor partly determines the internal supply voltage drops and, hence, the output voltage. So, when either the load current increases or the a-c input voltage decreases, the supply output voltage and, therefore, the voltage across R, tend to decrease—less current flows through R in accordance with Ohm's Law. As a result, the portion of the internal voltage drop in the supply caused by R goes down somewhat to increase the output voltage by the same amount.

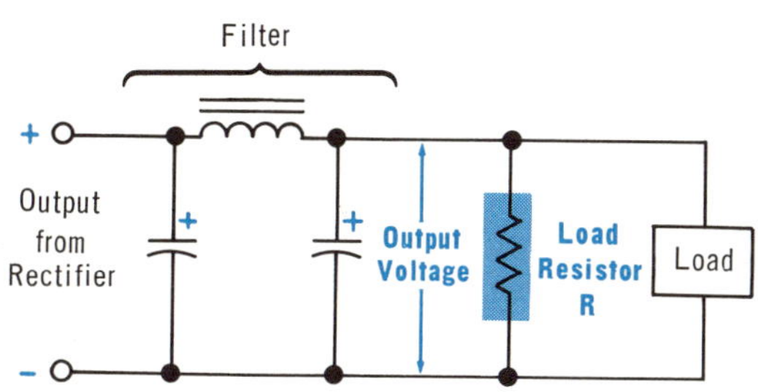

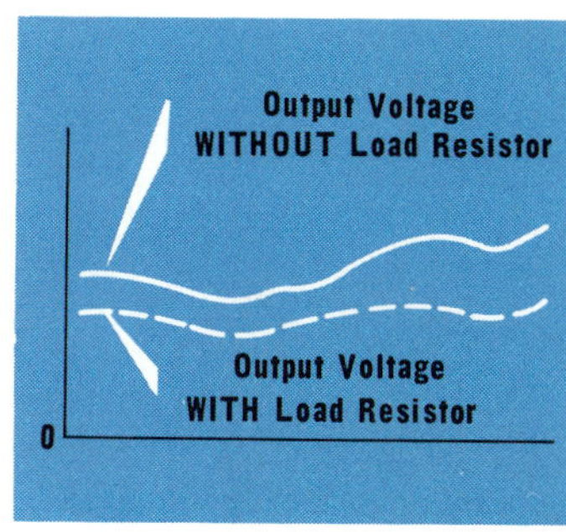

The load resistor decreases the average output voltage from the supply. It also provides a small degree of voltage regulation

The opposite situation occurs if the output voltage tends to increase. The resulting higher voltage causes more current to flow through the load resistor, which increases the internal voltage drops of the supply so that the output voltage goes down.

As was mentioned, the load resistor provides only a small degree of regulation. This is because the fluctuating output voltages only cause small changes in the load resistor currents. These small changes are not great enough to vary the internal voltage drops of the supply to bring the output voltage back to normal.

The load resistor regulator is sufficient with supplies where the line and load do not vary much. This regulator is most efficient when the load resistor draws 10 percent of what the load draws.

the gas-tube regulator

If the load resistor is replaced by a device that would cause large changes in internal voltage drops for small changes in output voltage, much better regulation could be obtained. Such a device could not follow Ohm's Law, since as the voltage across it went up or down, its resistance would have to go down or up to cause a higher than usual current change. Such a device is called *nonlinear*.

One such type of voltage regulator uses gas-filled, glow-discharge tubes, whose resistance is determined by the degree of ionization of the gas in the tube, which in turn depends on the current. The higher the current through the tube, the greater is the ionization, and the lower is the resistance. Theoretically, the resistance drops in proportion with any current increase so that the *IR* drop across the tube is constant. But in actual use, the series resistor is needed to bring this about.

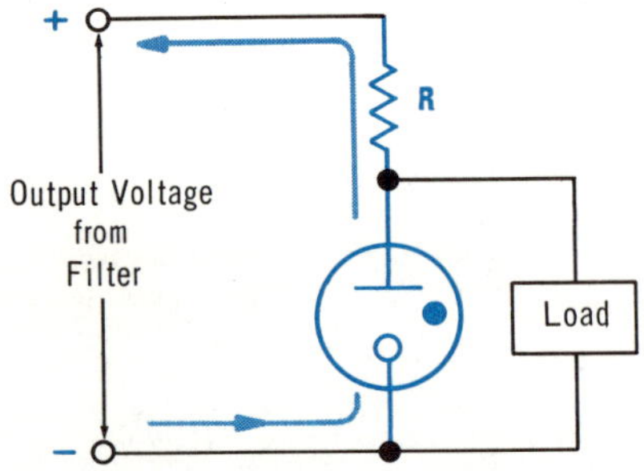

Changes in supply voltage appear across dropping resistor R. The voltage across the glow tube, and, therefore, the load voltage, remain constant

Both the load current and the glow-tube current flow through dropping resistor R. Part of the output voltage of the supply is dropped across resistor R, and the remaining voltage, which is the regulated output, is dropped across the gas tube. When the voltage across the tube tends to go up, more current flows through the tube. Gas ionization, then, increases, and the resistance of the tube goes down. As a result of this decrease in resistance, then, an even greater current flows through the tube. This would tend to increase ionization again, and reduce resistance to increase current again, but a stabilizing point is reached where the initial increase in voltage across the tube results in a much higher increase in current than could be obtained with a resistor. As a result, the voltage dropped across resistor R goes up sufficiently to lower the voltage across the tube to normal. Of course, the opposite occurs when the voltage across the tube tends to go down.

The value of dropping resistor R is very important, since it must match the current changes to pick up all of the voltage fluctuations. Also, the tube cannot give perfect regulation, for if the voltage returned to its exact value, the ionization within the tube would do so, also, and the action would repeat itself, causing the output voltage to fluctuate up and down. So, the regulated output voltage is only returned *close* to its regulated value for stabilization purposes.

series gas-tube regulators

Gas tubes are rated according to the constant voltage that they are designed to drop, as well as the maximum current that they can carry. For example, the VR 75-30 maintains a relatively constant 75-volt output and can carry up to 30 milliamperes. Various type tubes are available for different voltages. Sometimes, a larger supply output voltage must be regulated than is possible with the characteristics of a *single* tube. Two or more tubes can then be connected in *series*, and the regulated output taken from across the combination. Such an arrangement is shown in A. The operation of the circuit is essentially the same as that of a single-tube regulator.

A series arrangement of glow tubes can also be used to provide more than one regulated output voltage. In this case, different loads are connected across the various tubes in the series arrangement. Such a circuit is shown in B.

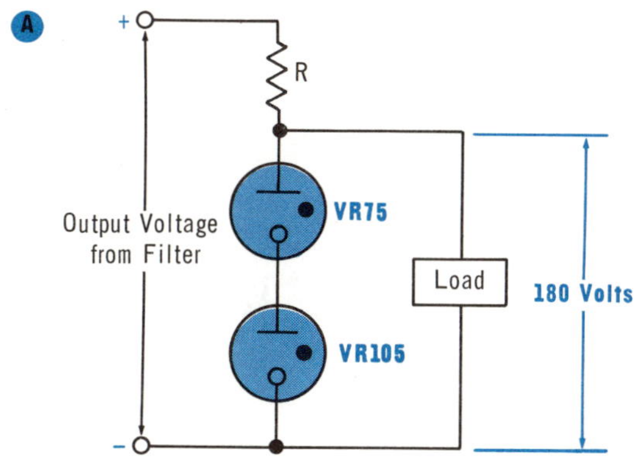

Gas-type regulator tubes can be connected in series, as well as singly

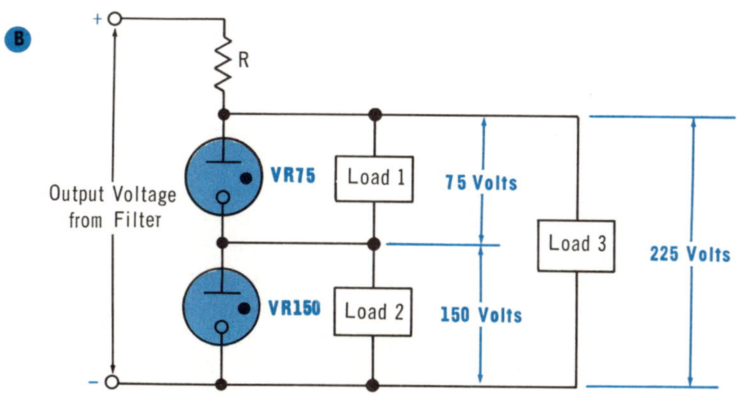

semiconductor regulators

Various semiconductor devices have certain nonlinear characteristics which, like the gas tube, make them suitable for voltage regulation purposes. Regulator circuits using two of these devices, the *thermistor* and the *zener diode*, are shown.

The thermistor, shown in A, has an extreme *negative temperature coefficient*. When current through it increases, the temperature is raised, and the resistance of the thermistor decreases. The thermistor is connected in series with a dropping resistor across the filter output. When the supply voltage rises, the voltage across the thermistor goes up, and so does the current through it. The higher current causes an increase in temperature that causes the resistance of the thermistor to decrease. This increases the current through the thermistor still further, so more voltage is dropped across the dropping resistor, and the output voltage reduces back to normal.

The thermistor regulator is only effective against *slow* changes in the supply voltage because it takes time for the temperature, and, therefore, the resistance, of the thermistor to change with the current variations. Another disadvantage of the thermistor regulator is that it can be affected by changes in temperature of the surrounding air.

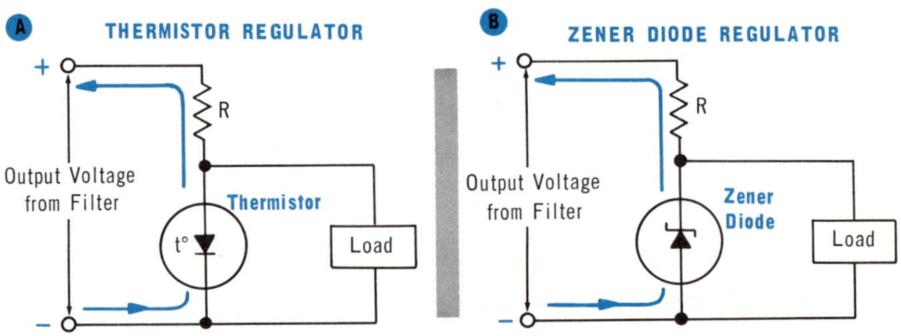

With a suitable value of dropping resistor (R), both the thermistor and the zener diode provide a constant voltage to the load despite changes in the output voltage from the filter

The zener diode regulator, shown in B, operates similarly to the thermistor regulator. The diode is reverse biased to operate on the zener breakdown region of its characteristic curve. In this region, the diode exhibits a wide nonlinear characteristic over a range of currents. When the output voltage falls, current through the diode decreases, causing an increase in the diode's effective resistance. This decreases the current still further, so that much less voltage is dropped across the series resistor, and the output voltage goes up to normal.

The zener diode regulator acts almost instantly and is not affected by ambient temperature.

summary

☐ The vibrator power supply produces a high value of dc from a low-voltage d-c source. The heart of a vibrator supply is a vibrating reed that chops the d-c input voltage. ☐ A nonsynchronous vibrator supply changes the d-c input voltage to ac, which is then stepped up by a step-up transformer and applied to a rectifier and filter. ☐ A synchronous vibrator power supply does not require a separate rectifier. It produces a stepped-up fluctuating d-c output that is then smoothed by a filter.

☐ In a multiphase or polyphase supply, the variously phased voltages are rectified separately and developed across a common output. ☐ The ripple frequency of a multiphase supply is proportional to the number of phases of the input. The higher the ripple frequency, the easier it is to filter.

☐ The load regulation of a power supply is the degree to which its output voltage remains constant despite changes in the amount of current drawn by the load. Voltage regulating circuits are used to keep the supply output voltage constant despite changes in load current. ☐ Load resistor regulation is most efficient when the load resistor draws only up to 10 percent of what the load draws. ☐ The gas tube regulator acts as a nonlinear device that tends to maintain a constant voltage across itself in spite of current changes through it. In this way it keeps the voltage applied to the load relatively constant. ☐ Specific voltage regulator tubes have a rated voltage that they are designed to drop, as well as a maximum current that they can safely carry. ☐ Gas regulator tubes can be arranged in series to provide for regulation of a greater voltage than is possible with a single tube. ☐ Certain semiconductor devices also have nonlinear characteristics that make them suitable as voltage regulators. Two such devices are the thermistor and the zener diode.

review questions

1. Draw a diagram of a nonsynchronous vibrator supply.
2. Explain the operation of a synchronous vibrator.
3. Does a synchronous vibrator require a transformer to operate?
4. What is meant by *regulation of a power supply*?
5. How does the thermistor voltage regulator work?
6. Draw a gas tube regulator.
7. Why is the series resistor needed?
8. How does a zener diode provide voltage regulation?
9. Why is a multiphase power supply used?
10. Draw a diagram showing how three loads can be regulated with only two glow tubes.

vacuum-tube voltage regulators

By using a vacuum tube as the voltage-regulating element, a much higher degree of regulation is possible than with the circuits previously described. This is because of the *amplifying* characteristic of a vacuum tube. Even very small voltage changes can be eliminated, since they are amplified sufficiently to cause the regulator circuit to operate better.

A simple vacuum-tube regulator circuit is shown in A. Triode V_1 acts as a variable resistance in series with the load. Battery B provides a bias for V_1 such that the tube has the proper plate resistance to drop the right amount of voltage to get the required output voltage. It is important to realize that the bias battery (B) voltage is *independent* of the supply output voltage. It remains constant regardless of how the supply voltage varies and serves as a *reference voltage*.

Vacuum-tube regulator circuits provide a very high degree of regulation. The tube acts like a variable resistor with gain to vary its voltage drop and keep the load voltage constant

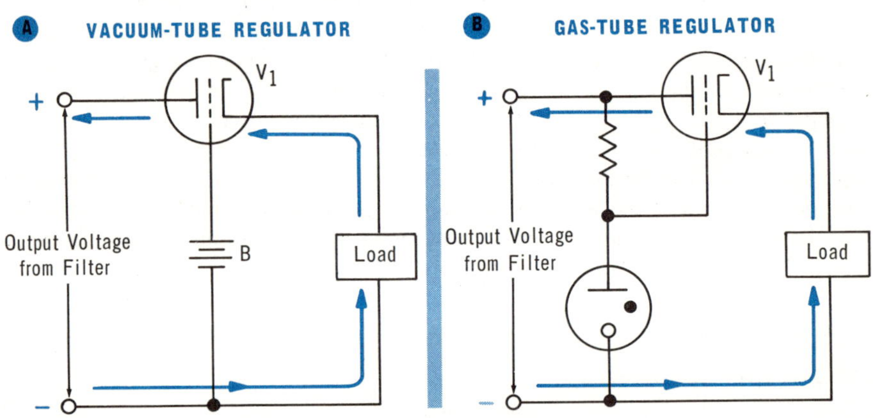

The gas-tube regulator is needed to supply a stabilized reference source for the bias

If the supply voltage increases, the cathode voltage of V_1, which is also the load voltage, tends to become more positive. This causes the negative bias on V_1 to increase also, since the grid battery voltage does not change. Consequently, the effective plate resistance of V_1 increases, as does the *IR* drop across V_1. Because of the tube's gain, the plate resistance increases by an amount sufficient to counteract the supply output voltage increase, so that the load voltage is returned to an essentially constant output. Because of the voltages involved, it usually is inconvenient to use batteries for the grid voltage. A gas tube with the proper voltage characteristic is used for this purpose. Its operation is the same as the gas-regulator tubes previously discussed.

the amplified voltage regulator

Because of the high amplification of the voltage fluctuations, this circuit is capable of compensating for even extremely small changes in output voltage

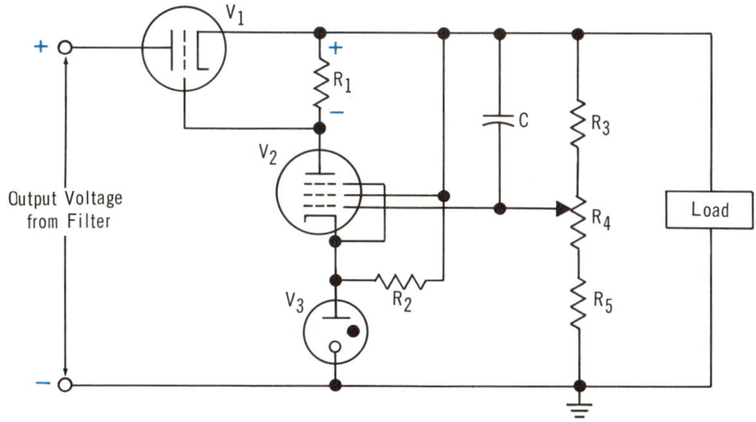

In the simple tube regulator, the change in plate resistance for a given change in bias is still not enough to give perfect regulation. However, it can be improved by first amplifying the voltage change.

In the circuit, triode V_1 and resistors R_3, R_4, and R_5 make up a voltage divider across the filter output. The resistor portion of the divider is in parallel with the load, so the load voltage appears across it. Pentode V_2 senses changes in the supply voltage, amplifies them, and varies the bias of V_1 to compensate for the changes. The cathode of V_2 is kept at a constant reference potential by gas regulator tube V_3. The control grid voltage of V_2 is tapped off of resistor R_4 and is *less positive* than the cathode voltage. It is initially adjusted to establish proper operation under normal load conditions. It accomplishes this by setting the bias of V_2 so that its plate current will drop enough voltage across R_1 to bias triode V_1 properly.

Because of the high gain of this regulator circuit, the divider resistor values are chosen to tap down the output voltage enough so that after the change is amplified, it will exactly cancel the output change.

Capacitor C increases the *response time* of the regulator in compensating for extremely *rapid changes* in supply voltage. The a-c component of these spurious voltage variations is coupled directly to the control grid of V_2 by the capacitor, which has a very low reactance at the high frequencies represented by sudden transient voltage changes. Capacitor C also eliminates any *ripple voltage* that still exists after filtering. It does this by applying the ripple voltage variations directly to the grid of V_2. The regulator thus compensates for these variations even more than it does for any other voltage changes, since these changes are not tapped down first.

transistor voltage regulators

A basic transistor regulator is shown in A. The transistor is in series with the supply line, and its voltage drop is varied to keep the output voltage constant. The base is held at the fixed reference voltage, supplied here by a zener diode regulator. The output voltage of the supply establishes the emitter voltage, and the difference between the base and emitter voltages provides the base-emitter bias, determining the resistance of the transistor. With the negative supply shown in A, the output voltage at the emitter is made slightly less negative than the base, since a p-n-p transistor is used. So, if the output voltage goes up, the emitter goes more negative, and there is less base-emitter bias. The resistance of the transistor, then, goes up to drop the output voltage back to normal.

A

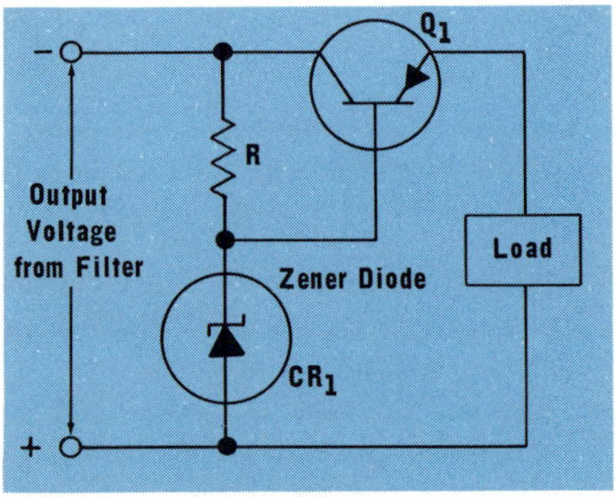

Transistor regulators perform the same function as their electron-tube counterparts

A transistor-amplifier regulator, comparable in operation to the tube-amplifier regulator of the previous page, is shown in B. Q_2 amplifies the tapped-down portion of the output voltage variations and varies the drop across R_1 to change the base voltage of Q_1. This varies the base-emitter bias on Q_1, so that the resistance, and, hence, the voltage drop across Q_1, will change to keep the output voltage constant.

B

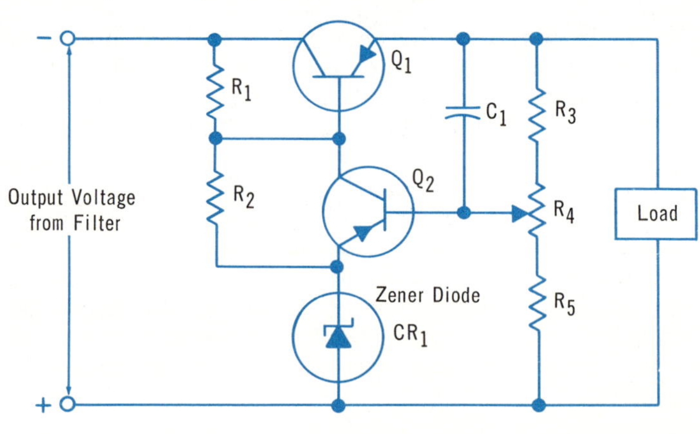

line regulators

Regulation can also be accomplished by increasing or decreasing the level of the *a-c input* to the supply to compensate for changes in supply output voltage. This is called a-c *line regulation.* It is accomplished by using a *saturable reactor* in series with the primary winding of the power transformer. If the impedance of the reactor is changed, the reactor will drop more or less voltage, and will allow less or more voltage to be applied to the power transformer. The impedance of the saturable reactor is changed by a d-c current from an amplifier that controls the degree to which the reactor is saturated. In the line regulator shown in A, a reduced output voltage will lower the bias to the amplifier, which then provides more saturating current for the saturable reactor. The more the reactor is saturated, the less inductance it has, the less a-c voltage it drops, the more ac is applied to the transformer, and the higher is the power-supply output voltage.

Often, a line regulator is used to compensate only for changes in the a-c input level. This is done, as in B, by using a second secondary winding on the power transformer to apply a portion of the a-c voltage to a separate rectifier, which feeds a d-c output to the control amplifier.

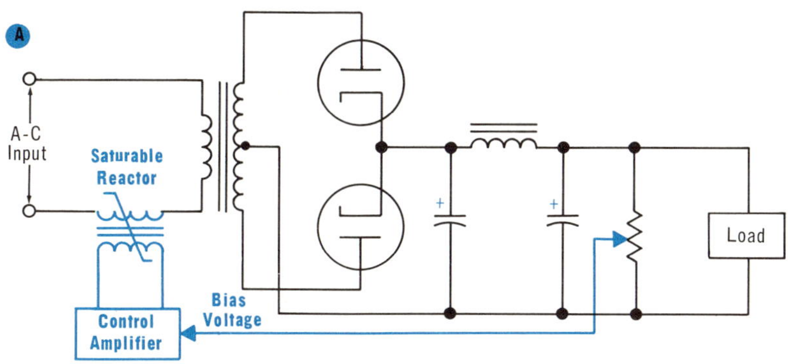

In this circuit, the a-c input voltage is varied to compensate for changes in the d-c output voltage. If the output voltage goes up, the bias to the amplifier will go up to reduce the saturating current. The impedance of the saturable reactor will go up to reduce the primary voltage, and, hence, the supply output voltage

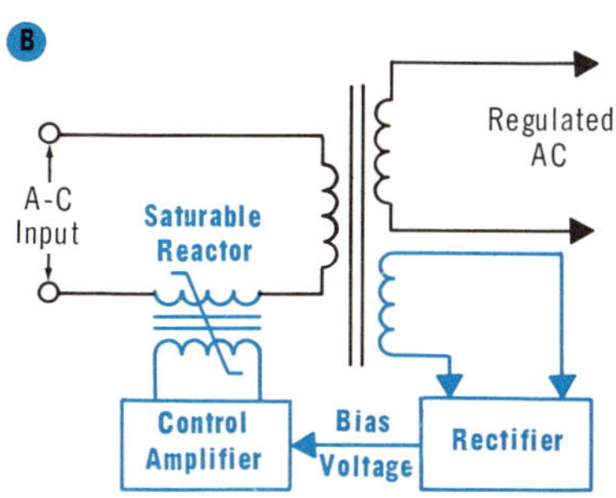

This line regulator maintains a constant a-c output from the power transformer regardless of variations in the a-c input

current regulation

Sometimes a power supply must deliver a *constant current* regardless of changes in supply voltage or load. This can be done utilizing the circuit shown.

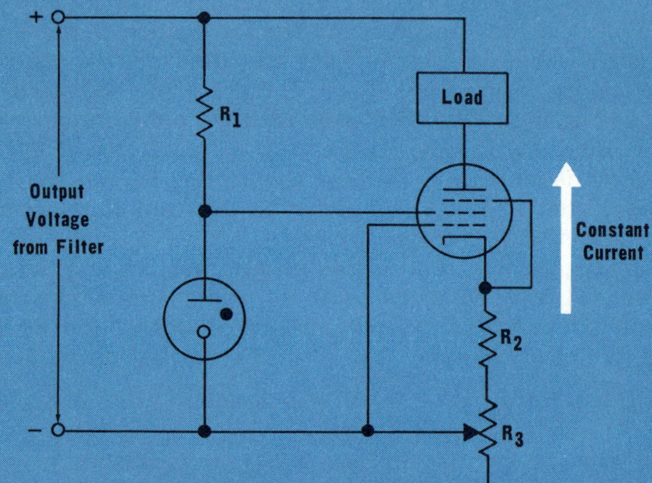

A current-regulator circuit maintains the output current from a power supply constant

A pentode is connected in series with the load across the supply output. So, the supply output current must flow through the pentode. Two factors cause the pentode to pass a relatively constant current: (1) the plate resistance of the pentode is set by its bias to be high compared to the resistance of the load, so that any changes in load resistance have less effect on the load current drawn; (2) the manner in which the pentode's bias is developed, an explanation of which follows.

The control grid is connected directly to the negative side of the supply and the cathode is connected through resistors R_2 and R_3 to the same point. The load current passes through these resistors. If the load resistance drops to draw more current, the drops across R_2 and R_3 will go up to increase the tube bias and raise the tube resistance. The tube resistance will go up as much as the load resistance went down, so that the total resistance across the supply remains constant to keep the load current constant.

The screen grid in this circuit is held at a constant voltage by a gas regulator tube. This is necessary to prevent fluctuations in the screen voltage, which would affect the plate current. Resistor R_3 is set for the desired load current level.

the complete power supply

You have now seen the various circuits that make up an electronic power supply. The different types and combinations of these circuits that are used in a particular power supply depend on the requirements of the supply. A power supply can be as simple as that below, consisting merely of a half-wave rectifier and a capacitor filter, or it can be as complete as that shown on the following page.

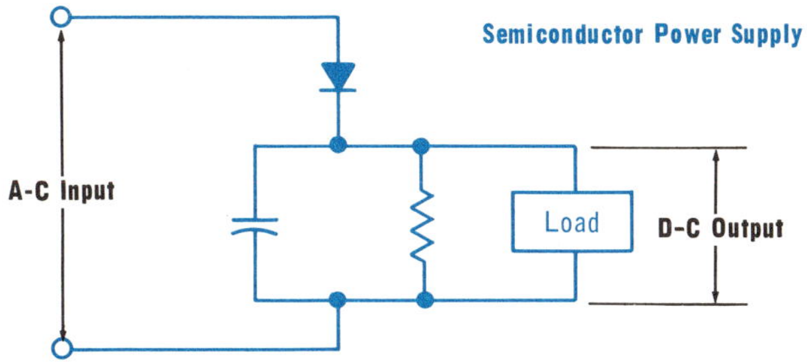

An extremely simple power supply consists only of a half-wave rectifier and a capacitor filter

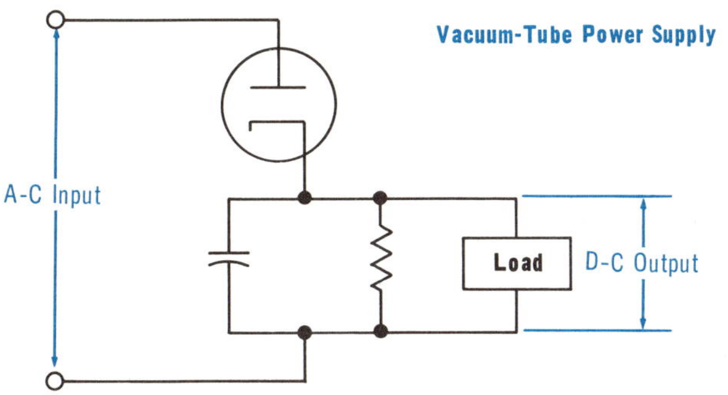

In the supply shown on the next page, a power transformer first steps up the a-c input voltage to a considerably higher value. The transformer secondary is centertapped, and the voltages across the two halves are applied to the diodes of a full-wave rectifier. The diodes conduct on alternate half cycles of the a-c voltage and deliver a fluctuating d-c output to a capacitor-input pi filter.

In the filter, the fluctuating dc is smoothed by the charging and discharging of the capacitors and the choke's opposition to current changes. The rectified and filtered voltage is then held relatively constant by a vacuum-tube voltage regulator. The load, therefore, receives a constant voltage. A bleeder resistor is used across the output of the supply to provide a discharge path for the filter capacitors when the supply is turned off.

the complete full-wave rectifier

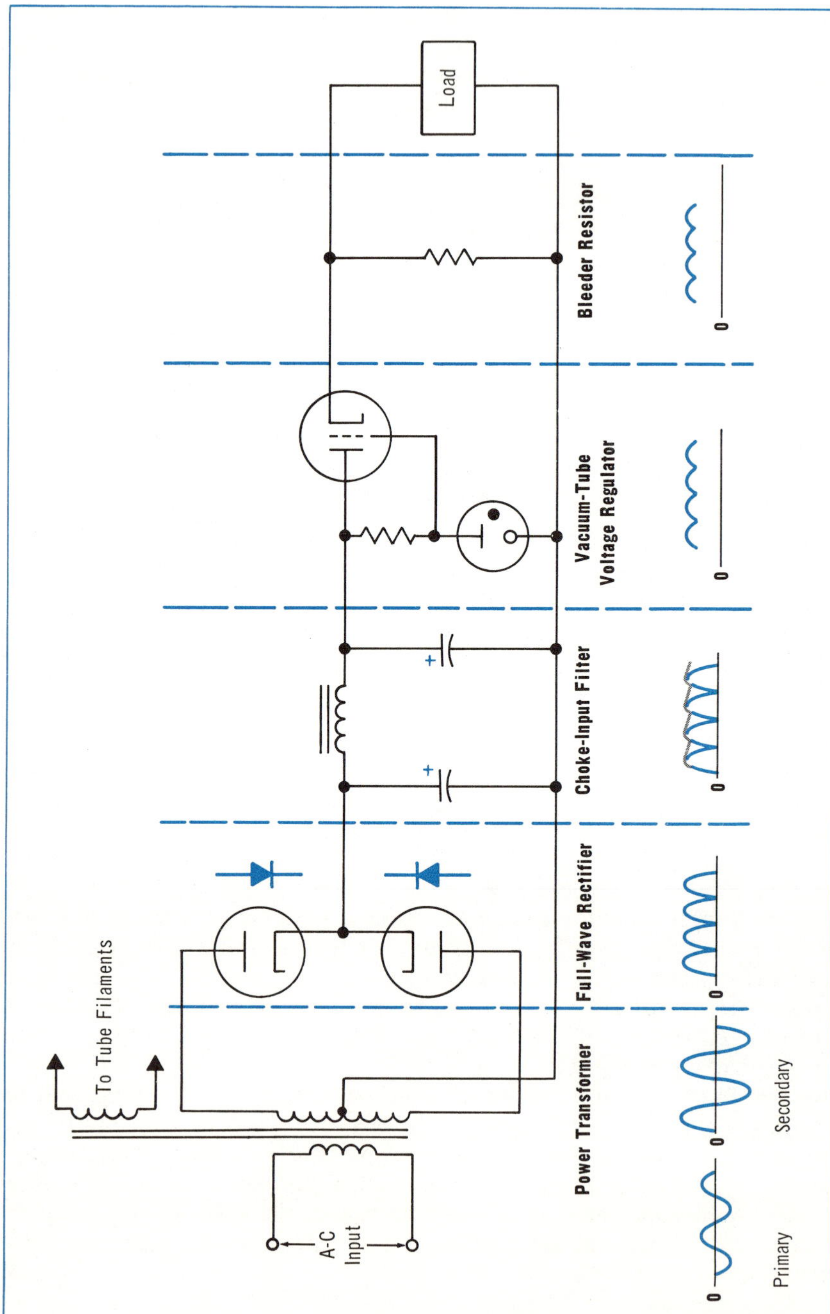

power supply characteristics

You have now learned about the various types of circuits that make up a power supply, the functions of these circuits, and the way in which they operate. All of the individual circuits of any given power supply function together to give that supply its overall operating characteristics. These characteristics then define the capability of the supply. From a functional standpoint, a power supply can be fully described on the basis of the characteristics given below.

POWER SUPPLY CHARACTERISTICS

Characteristic	Description
Input	The a-c input power that must be applied to the supply for it to deliver its rated output. It includes the required amplitude, frequency, and phase. For example, a typical power supply might require an input of 110 volts, 60 cps, single phase.
Output	The rated voltage output of the supply, together with the maximum current that can be drawn without affecting the voltage. If a supply produces more than one output voltage, its rated output consists of each of these voltages and their associated maximum currents. For example, the output of a typical supply might be +250 volts dc at 50 milliamperes and −80 volts dc at 100 milliamperes.
Percent Regulation	The percentage decrease in the supply's output voltage from no-load to full-load conditions. The higher the percentage of regulation, the more the output voltage falls as the load is increased. Thus, if a supply had zero percent regulation, its output voltage would remain absolutely constant from no load to full load.
Percent Ripple	The amount of ripple voltage that is not removed by the filter, and which therefore appears at the output of the supply. The lower the percentage of the ripple, the more nearly the output voltage approaches pure dc.

summary

☐ Vacuum tubes or transistors, because of their ability to amplify, can provide a very high degree of regulation. ☐ In an amplified voltage regulator, changes in the supply output voltage are amplified before being applied to the regulating portion of the voltage regulator circuit. This makes the circuit capable of compensating for even extremely small changes in output voltage. ☐ The operating point of a tube or transistor regulator is set by a reference voltage that is independent of the supply output voltage. Gas regulator tubes or zener diodes are used to establish the reference voltage.

☐ Regulation of a supply output voltage can also be accomplished by increasing or decreasing the level of a-c input to the supply. This is called a-c line regulation. ☐ In a-c line regulation, a saturable reactor is used to change the impedance of the power transformer primary winding in accordance with changes in the supply output voltage. ☐ A line regulator can also be used to compensate for changes in the a-c input level to the supply.

☐ Current regulation is used when a power supply must deliver a constant output current regardless of changes in line or load voltage. ☐ A pentode, connected in series with the load, is one method used for current regulation. The high plate resistance of the pentode causes it to act as a constant-current source. ☐ The choice of a power supply for a particular application is determined by the requirements of the situation. Some important characteristics of a supply are input, output, percent regulation, and percent ripple.

review questions

1. Why is an amplified voltage regulator an improvement over a simple regulator?
2. What is meant by the *response time* of a voltage regulator?
3. Why are pentodes used for current regulation?
4. Are pentodes ever used for voltage regulation?
5. What is a *saturable reactor*?
6. Name three important characteristics of a power supply.
7. What device is used in an a-c line regulator?
8. What is the difference between an a-c line regulator and a regular voltage regulator?
9. Draw a schematic diagram of a simple transistor voltage regulator. Describe its operation.
10. What is meant by the *percent ripple* of a power supply?

amplifier circuits

In practically every piece of electronic equipment there are at least one or two, and, often, many circuits whose only function is to *increase* the *level* of the signal. These *amplifier circuits* may use electron tubes or transistors, since both are essentially amplifying devices. Besides the tube or transistor, an amplifier circuit includes all of the circuit components that provide *bias*, as well as any components used to *compensate* for undesired *distortion* of the signal that occurs during amplification.

Sometimes, an amplifier is designed *purposefully* to distort the signal to obtain some desired waveform characteristic. When this is the case, components that contribute to the desired distortion are also part of the amplifier circuit. Very often, a signal is so small that one amplifier circuit cannot amplify it to the desired level. In these cases, two or more amplifier circuits are connected in *sequence*, with the amplified output of the first being used as the input to the second for further amplification, and so on.

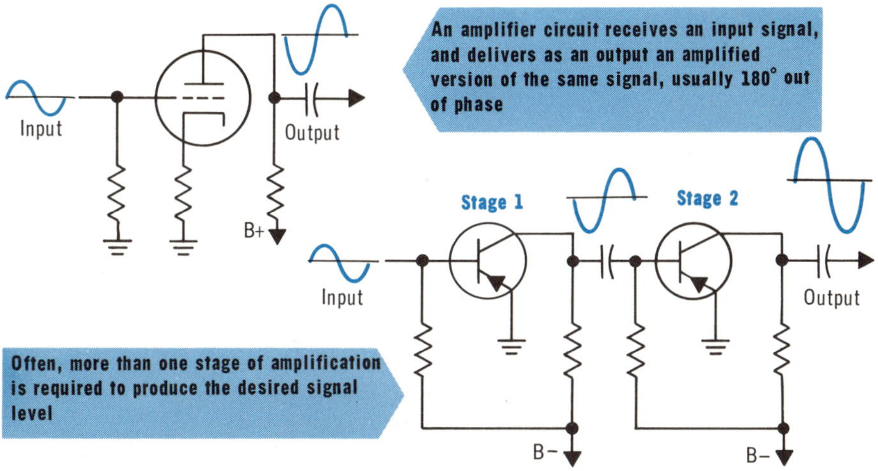

You recall from your previous studies that *coupling circuits* must then be used between the individual amplifier circuits. When more than one amplifier circuit is used, it is convenient to refer to each as a *stage*, and then consider the signal as having undergone one, two, three, etc., stages of amplification.

Amplifier circuits can be classified in a wide variety of ways. They can be classified according to their *use*, the type of *bias* used, the *frequency* or *bandwidth* of the signals they are to amplify, the type of *coupling*, if more than one stage is used, and their circuit *configuration*. The various types of amplifiers within each of these classifications have already been covered for the most part in previous volumes; they will therefore be summarized on the following pages.

amplifiers according to use

When amplifiers are classified according to their use, they fall into two broad groups: *voltage amplifiers* and *power amplifiers*. Voltage amplifiers, as their name implies, increase the voltage level of an applied signal. Since the output voltage of an amplifier is determined by the voltage drop across the output load, the *impedance* of the load is made as *large* as is practical in most voltage amplifiers.

Power amplifiers are designed to deliver a large amount of current to the output load. To accomplish this, the load impedance is usually *low enough* to allow a high current output, but at the same time it is not so low that it produces excessive distortion of the signal. Power amplifiers are also called *current amplifiers*.

Some other classifications of amplifiers according to use are the *buffer* amplifier, which is used to isolate other stages, the *square-wave* amplifier, which is designed to handle a specific waveshape, and the *frequency-doubler* amplifier, which increases the signal frequency. There are many other such amplifiers, which you will learn about later.

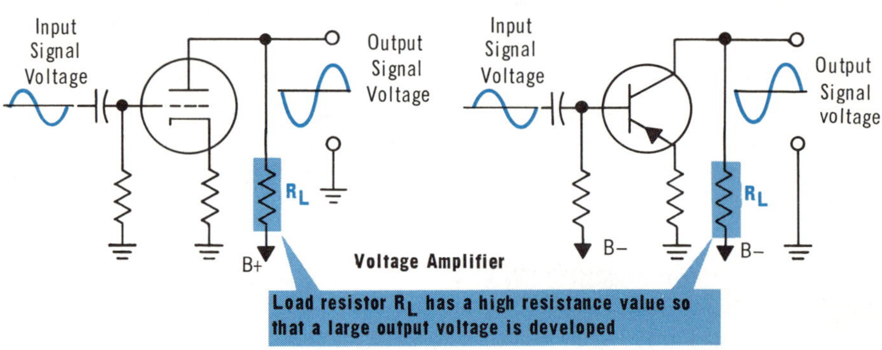

Voltage Amplifier

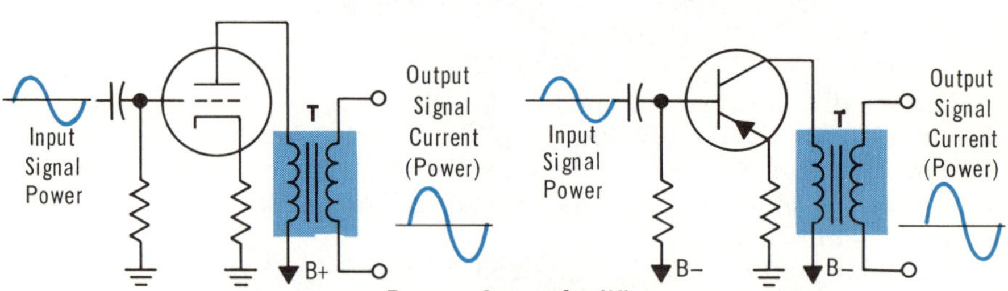

Power or Current Amplifier

The load resistance is a complex function of the transformer (T) turns ratio and resistance of the transformer secondary winding. It is relatively low so that sufficient signal current is developed in the transformer secondary

amplifiers according to bias

Amplifiers are also classified according to their biasing conditions, or, in other words, according to the portion of the input signal voltage cycle during which output current flows. There are four classes of amplifiers according to bias: class A, class B, class AB, and class C.

Class A amplifiers are biased in the center of their operating curves so that output current flows during the *entire* cycle of the input voltage. This results in *minimum distortion* of the output signal, and, as a result, class A amplifiers are widely used in audio systems, where low distortion is important.

Class B amplifiers are biased at cutoff so that output current flows for approximately *one-half* of the input-signal voltage cycle. When no input signal is present, no output current flows. In effect, a class B amplifier cuts off one half of the a-c input-signal waveform.

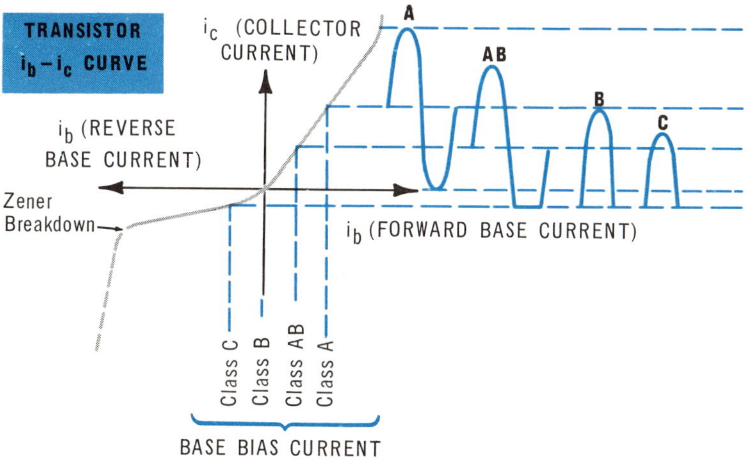

Transistor classes of operation differ somewhat from tubes because there is a reverse current beyond cutoff. Also, class C cannot be far beyond cutoff because the signal swing might drive the transistor into the zener breakdown region

Class AB amplifiers are biased so that output current flows for *appreciably more* than one half of the input cycle, but for less than the entire cycle. Essentially, class AB amplifiers are a compromise between the low distortion of class A amplifiers and the high efficiency of class B amplifiers.

Class C amplifiers are biased beyond cutoff so that output current only flows during the *positive-going peak* of the input cycle. Such amplifiers have high power outputs. They also have a high degree of distortion, which prevents their being used for audio applications.

The curves shown on the next page represent the various classes of bias for electron-tube circuits. Similar curves are developed for transistor circuits.

amplifiers according to bias (cont.)

CLASS A

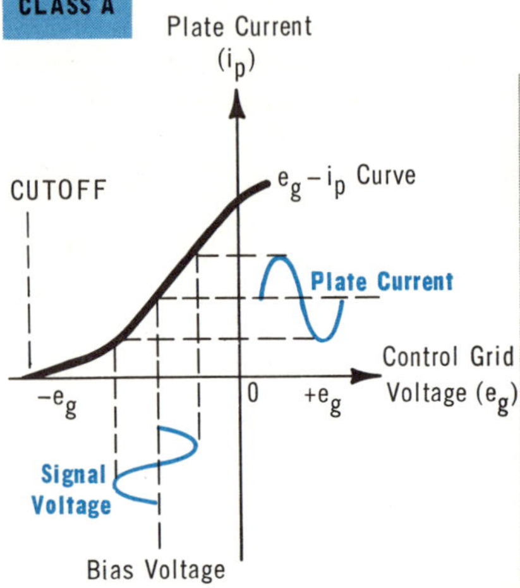

Plate current flows for 100 percent of input signal cycle

CLASS B

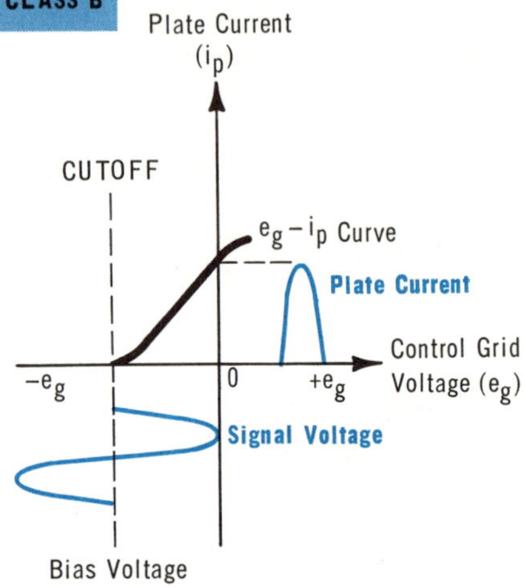

Plate current flows for approximately one half of input signal cycle

CLASS AB

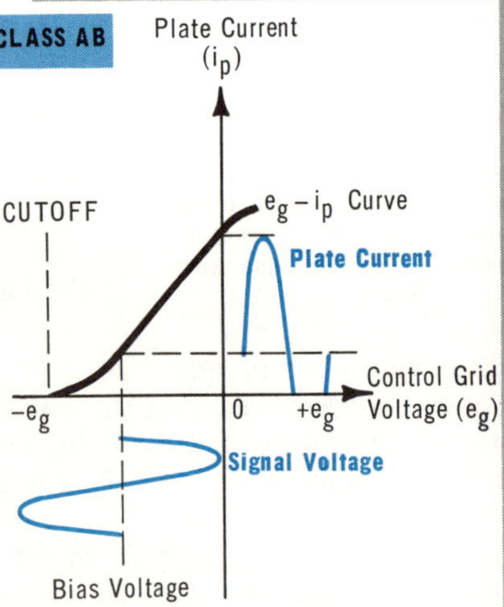

Plate current flows for more than half but less than 100 percent of signal cycle

CLASS C

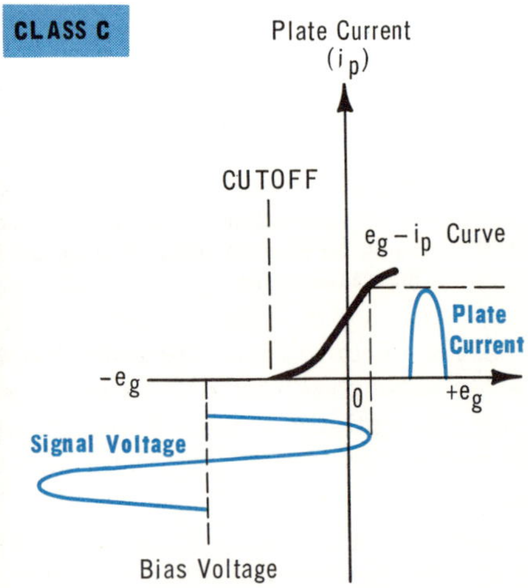

Plate current flows only for small portion of input signal cycle

amplifiers according to coupling

In applications where *more* than a single amplifier stage is used, the amplifiers are often classified according to the way in which they are coupled. The basic coupling methods are: *resistance-capacitance* (RC) coupling, *impedance* coupling, *transformer* coupling, and *direct* coupling. These are illustrated on the following page.

When amplifiers are RC coupled, the output load is a resistance, usually of relatively high value. RC-coupled amplifiers have good frequency-response characteristics over a relatively wide frequency range. Above and below this range, the gain falls off. The decrease in gain at the lower frequencies is caused by the increased reactance of the coupling capacitor. At the higher frequencies, the decrease in gain is caused by the decreased reactance of the interelectrode capacitances of the stages, as well as the stray capacitance of the wiring between stages. The decrease in these capacitances causes more of the signal voltage to be shunted, or bypassed, to ground.

In impedance coupling, a *coil* is used as the output load instead of a resistance. The frequency response of an impedance-coupled amplifier is not uniform, or flat, like that of an RC-coupled amplifier, because the impedance of the load coil varies with frequency, causing the amplifier voltage gain to increase with frequency up to the point where it is limited by the circuit shunting capacitances.

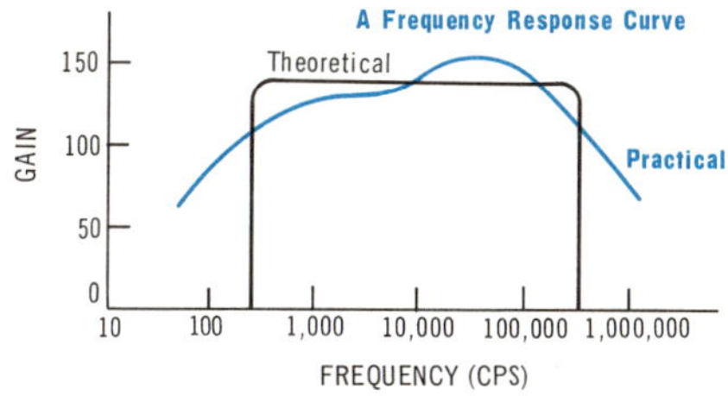

. . . is a graph of the gain of an amplifier plotted against frequency. A perfect response curve would be absolutely flat over the entire range of frequencies amplified. Such a response is impossible to achieve in actual circuits. One of the reasons for this is the reactive components, both physical and stray, in the coupling circuits

In transformer coupling, the output of one circuit is coupled to the input of the next circuit by means of a transformer. This can also provide additional amplification if the transformer has a step-up turns ratio. Transformer coupling, though, is expensive and bulky. It very often has poorer frequency response at the higher and lower audio frequencies than does an RC-coupled amplifier.

In direct coupling, the output of one stage is applied directly to the input of the next stage. A rather elaborate voltage distribution arrangement is required to provide the proper d-c operating voltages to amplifiers coupled in this way. When direct coupling uses *no reactive components,* it is especially suitable for amplifying low frequencies, as well as d-c signals. The absence of reactive components results in direct coupling having a frequency response that is uniform over a very wide range. However, d-c amplifiers are affected by voltage *drift.*

This has only been a brief summary. Detailed information on coupling was given in Volumes 3 and 4.

amplifiers according to coupling (cont.)

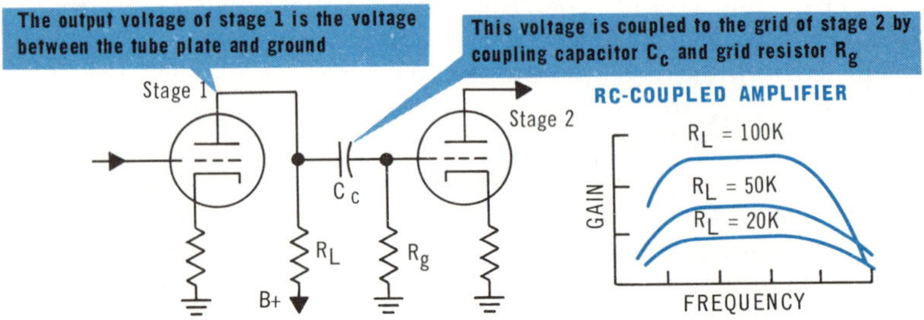

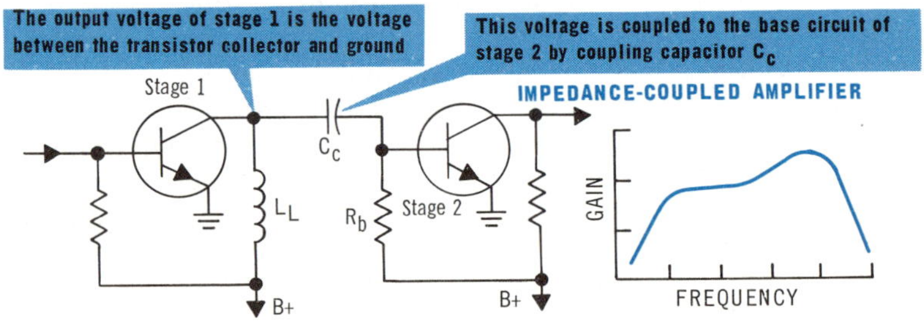

The response curve of an impedance-coupled amplifier is not as flat as that of an RC-coupled amplifier because the reactance of L_L and, therefore, the signal voltage, increases with frequency

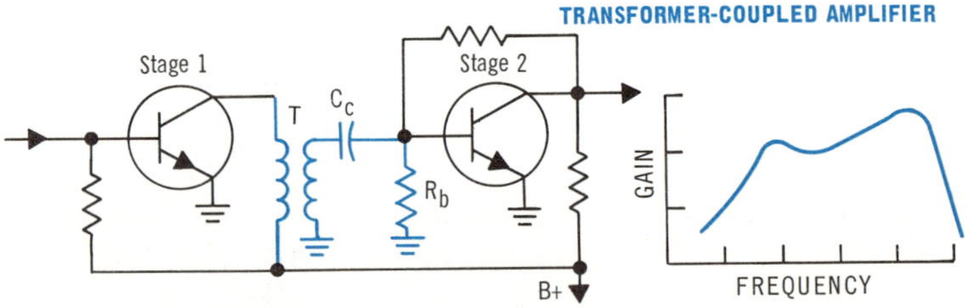

The output of stage 1 is coupled to the base circuit of stage 2 by transformer T, coupling capacitor C_c, and resistor R_b

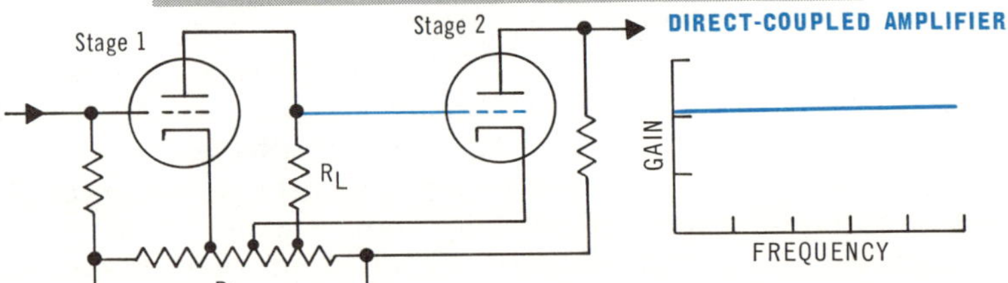

The output of stage 1 is connected directly to the input of stage 2 in a direct-coupled amplifier

amplifiers according to circuit configuration

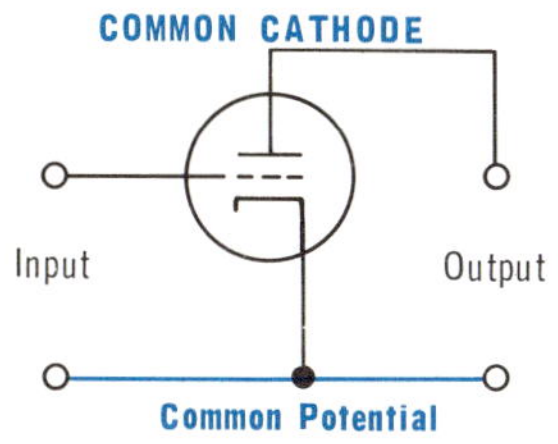

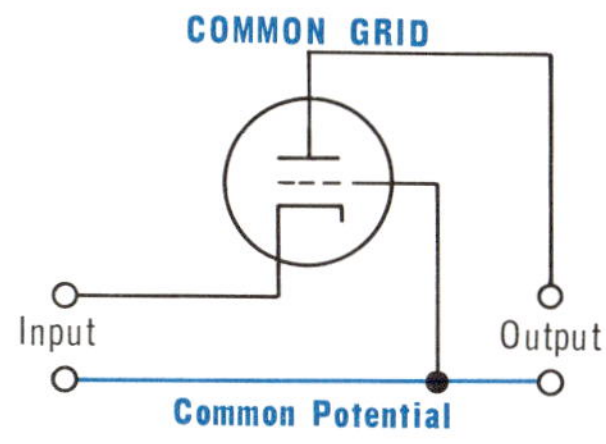

In many amplifier circuits, the input and output signal voltages are measured with respect to a common potential

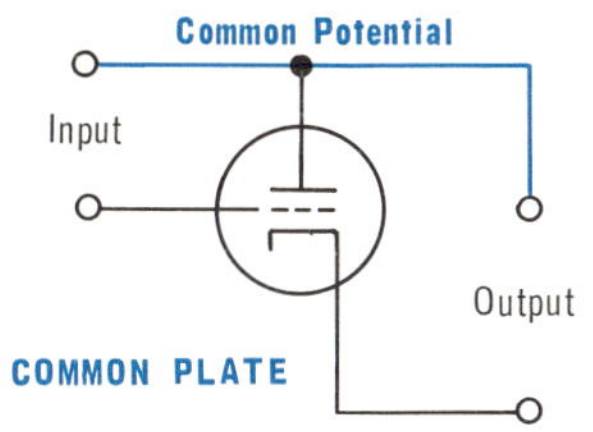

One element of the tube or transistor is connected to the common potential. This determines whether the circuit is of the common-grid, common-cathode, or common-plate type

The common potential is usually the zero reference for voltage measurements, and is designated as ground

Another way of classifying amplifiers is to specify the principal elements returned to *ground*. Thus, in amplifier circuits using electron tubes, there are grounded-cathode amplifiers, grounded-grid amplifiers, and grounded-plate amplifiers. Similarly, in transistor amplifier circuits, there are grounded-base amplifiers, grounded-emitter amplifiers, and grounded-collector amplifiers.

You should understand that an element does not have to be connected *directly* to ground to be considered grounded. It may be connected to ground through a component, such as a resistor or capacitor. The important point is that the potential of the element is at *a-c ground,* which is the *zero-potential reference point* for measuring all *signal* voltages, both input and output.

Another way of looking at this is to consider an amplifier as a *four-terminal circuit.* The input signal is applied to two of the terminals and the output is taken from the other two. However, one input terminal and one output terminal are usually *common* and are connected to the same tube or transistor element. You then have a common cathode, common emitter, and so forth, circuit. The common reference point is usually designated as ground.

The different amplifiers typed according to circuit configuration are shown on the following page.

amplifiers according to circuit configuration (cont.)

Grounded-Cathode Amplifier

The input is applied between the control grid and ground

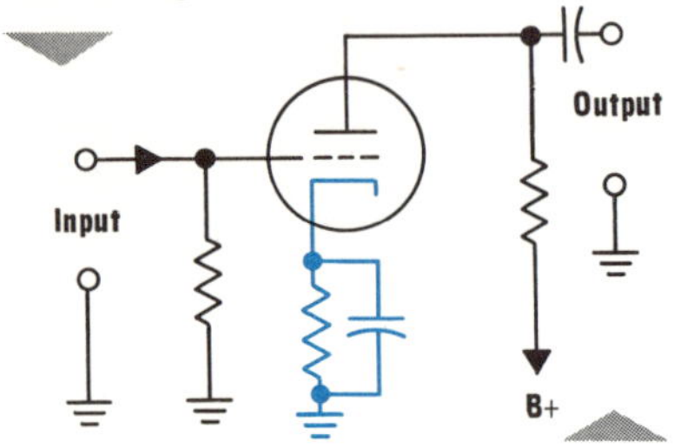

The output is taken from between the plate and ground

Grounded-Emitter Amplifier

The input is applied to the base-emitter circuit

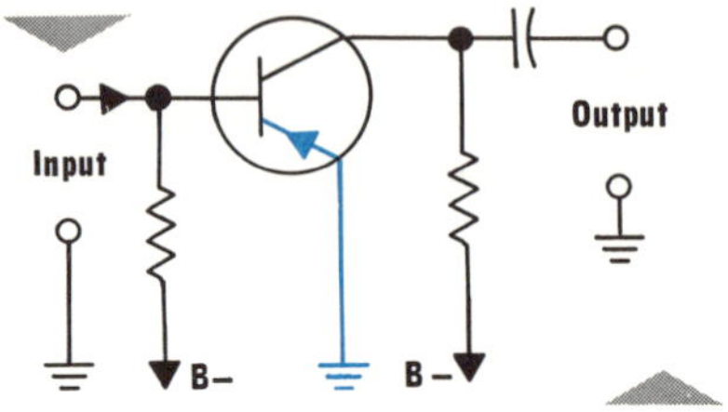

The output is taken from the collector-emitter circuit

Grounded-Grid Amplifier

The input is applied between the cathode and ground

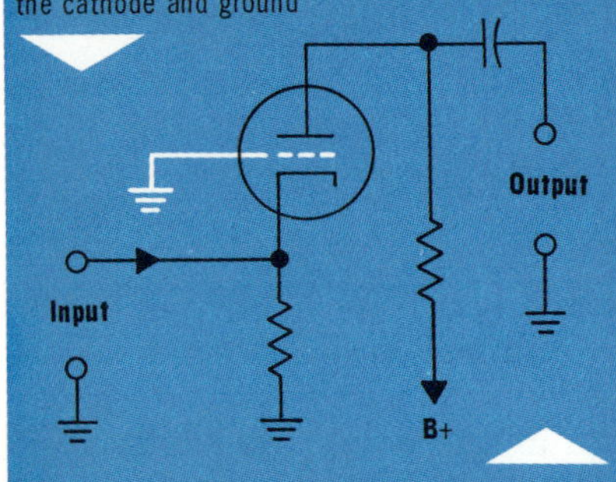

The output is taken from between the plate and ground

Grounded-Base Amplifier

The input is applied to the emitter-base circuit

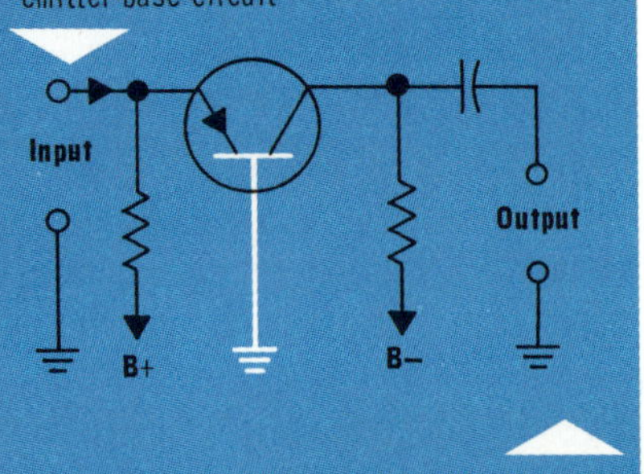

The output is taken from the collector-base circuit

Grounded-Plate Amplifier

The input is applied between the control grid and ground

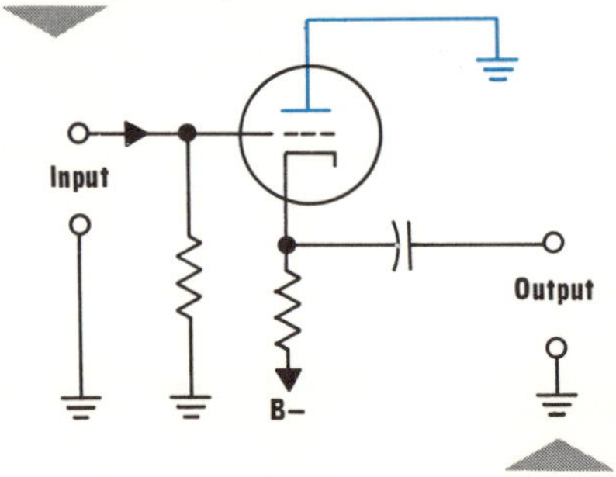

The output is taken from between the cathode and ground

Grounded-Collector Amplifier

The input is applied to the base-collector circuit

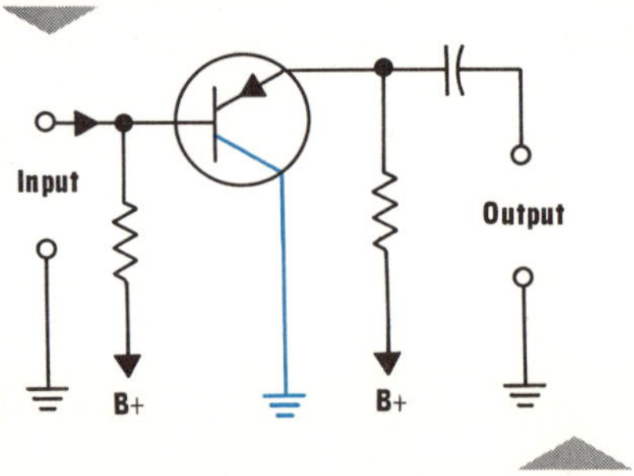

The output is taken from the emitter-collector circuit

amplifiers according to bandwidth

Sometimes, amplifiers are classified according to bandwidth. On this basis of classification, there are two principal types of amplifiers: those that are *tuned* and amplify a *restricted range* of frequencies; and those that are *untuned* and amplify a *wide range* of frequencies.

A tuned amplifier usually has a tuned (*resonant*) circuit in its input or output, or both, that passes only a relatively narrow band of frequencies. The center of this frequency band is the resonant frequency of the tuned circuit. The width of the band depends on the Q of the tuned circuit. In some tuned amplifiers, the band of frequencies that is amplified is *permanently fixed* by the values of the tuned-circuit components. In other tuned amplifiers, the frequency band that is amplified can be *varied* by changing the value of one of the components of the tuned circuit.

Tuned amplifiers use tuned circuits in their inputs, their outputs, or both

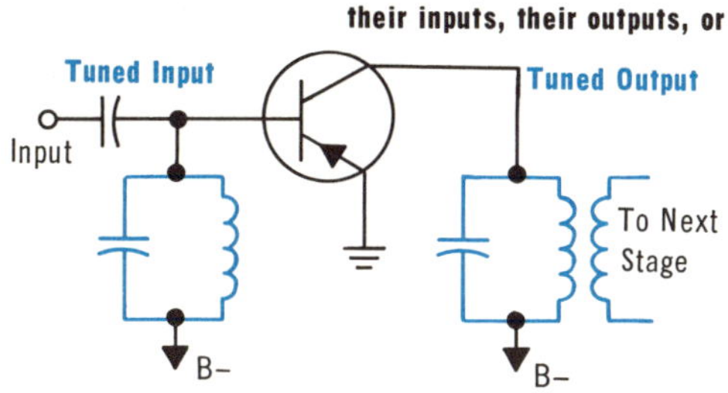

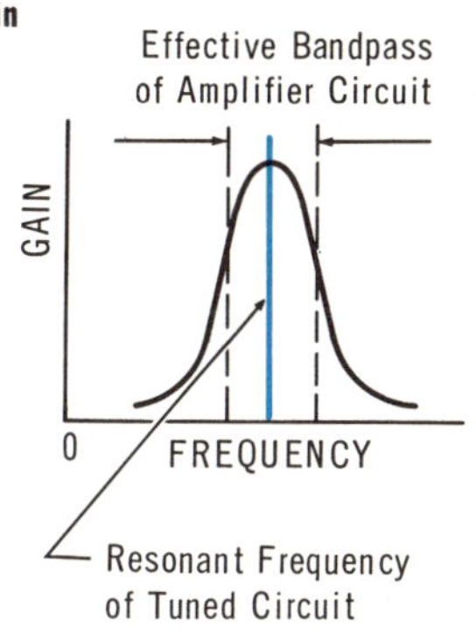

The tuned circuits restrict the range of frequencies that can be amplified by the amplifier to a relatively narrow band centered around the resonant frequency of the tuned circuit

Untuned amplifiers are not tuned to any specific band of frequencies. The range of frequencies that they can amplify, though, is still limited to some extent by the circuit components and stray capacitances. Compared to tuned amplifiers, however, they can amplify an extremely wide range of frequencies.

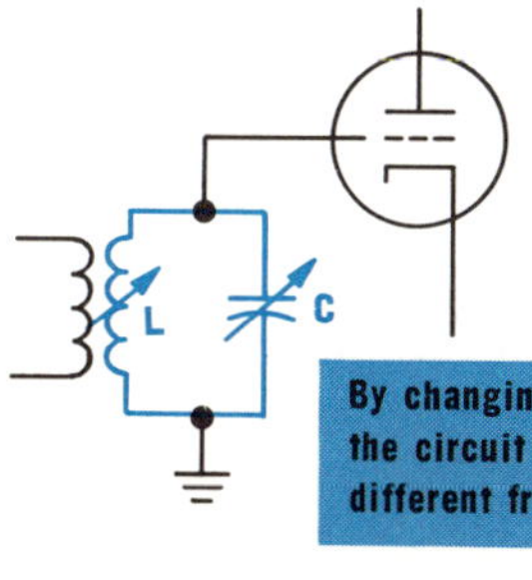

By changing the value of C or L, the circuit responds to different frequencies

In many tuned amplifiers, the resonant frequency of the tuned circuit can be varied. In this way, different bands of frequencies can be amplified

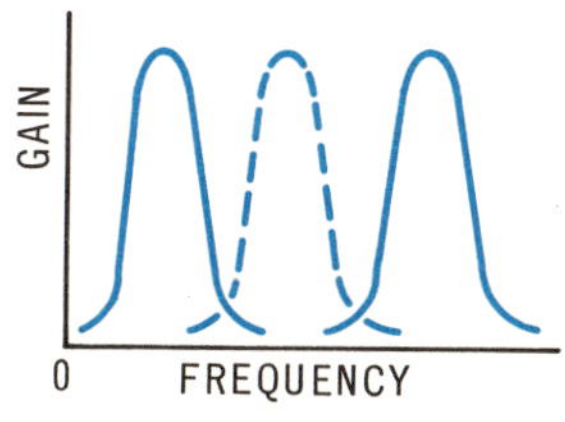

amplifiers according to frequency

The most common amplifier classification is according to *frequency.* This is especially useful in the description and analysis of circuits and equipments used for the radio transmission of electronic signals. According to frequency, amplifiers are classified as: *direct-current* (d-c) amplifiers, *audio-frequency* (a-f) amplifiers, *intermediate-frequency* (i-f) amplifiers, *radio-frequency* (r-f) amplifiers, and *video-frequency* amplifiers.

As their name implies, d-c amplifiers amplify d-c signals. Audio amplifiers operate in the audio-frequency range, or from about 20 to 20,000 cps. Video amplifiers are capable of amplifying signals from the lower audio frequencies to as high as 4 or 5 megacycles. You can see that video amplifiers are characterized by *very wide bandwidths.*

This Audio Amplifier is . . .

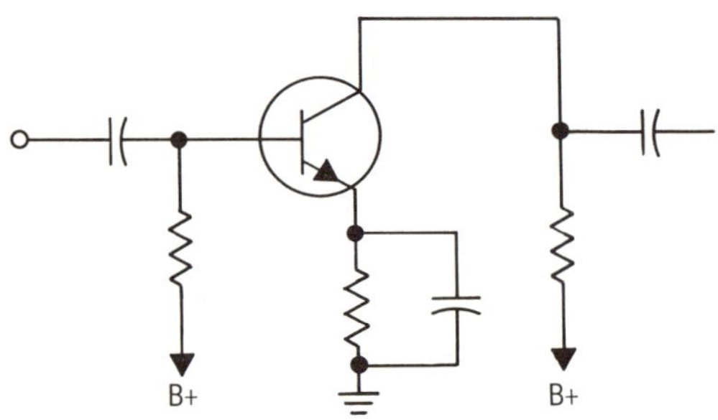

A Voltage Doubler
AND
A Class A Amplifier
AND
An RC-Coupled Amplifier
AND
A Grounded-Emitter Amplifier
AND
An Untuned Amplifier

Practically every amplifier circuit fits into all of the different classifications previously described

I-f and r-f amplifiers are not defined in terms of a specific frequency range. Instead, they are defined by the *nature* of the frequencies they amplify. Generally, they are *tuned* amplifiers, and therefore amplify a relatively small band of frequencies. I-f amplifiers operate at the intermediate frequency of a particular piece of equipment, and r-f amplifiers are tuned to the frequencies of various r-f carrier waves.

Later in this volume, amplifier circuits are broken down according to frequency (audio, video, r-f, and i-f), and described in detail. You will find that this method of grouping amplifiers is somewhat arbitrary, inasmuch as all of the methods of classification overlap to some degree. For instance, audio amplifiers may also be voltage or power amplifiers. Similarly, an r-f amplifier is usually a tuned amplifier, and at the same time may be a grounded-cathode or grounded-grid amplifier. In spite of this overlap, for purposes of description, some grouping must be used. From the standpoint of learning how amplifier circuits contribute to the overall operation of many types of electronic equipment, the best grouping is on the basis of frequency.

amplifier distortion

With the exception of certain waveshaping and special-purpose amplifiers, the output of an ideal amplifier is *identical* to the input except for an increase in amplitude, and, more often than not, a 180-degree phase reversal. In practical amplifiers, however, these ideal conditions are not met. Special distortion of the output signal always occurs.

Amplifier circuits are designed so that the degree of distortion is kept to a level that can be tolerated. Sometimes this can be accomplished by using one particular type of amplifier circuit rather than another. In other cases, though, it requires the use of additional components or circuits to compensate for distortion.

There are three main types of amplifier distortion: *frequency distortion, phase distortion,* and *amplitude distortion.* Frequency distortion occurs when some frequency components of a signal are amplified more than others. This type of distortion results because various circuit components, such as coupling capacitors, respond differently to different frequencies.

Phase distortion occurs because most coupling circuits *shift* the phase of a sine wave by an amount that depends on its frequency. Thus, each sine-wave component of a complex waveform is shifted in phase by a different amount. This causes distortion of the overall waveform. Phase distortion is not important in the processing of signals that represent voice or music, since the human ear cannot detect phase shifts of the individual components of a complex wave. In applications such as radar or television, though, phase distortion is of considerable importance.

Amplitude distortion occurs when equal changes in the amplitude of the amplifier input signal do not produce proportionate changes in the output. An extreme example of this is an input signal that drives an amplifier into *saturation.* As maximum output current flows at saturation, when the input increases further, the output amplitude does not change.

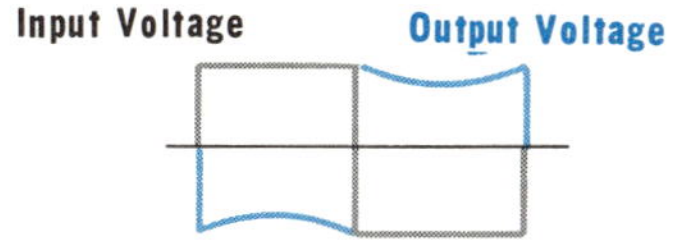

Frequency distortion occurs when some frequency components of a signal are amplified more than others

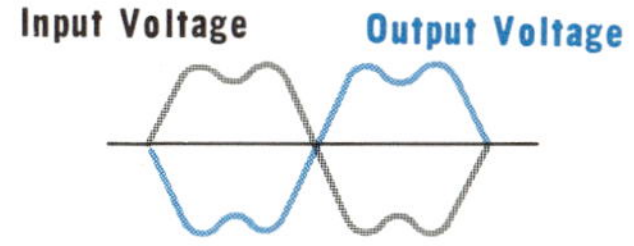

Phase distortion occurs when the various frequency components of a signal are shifted in phase by different amounts

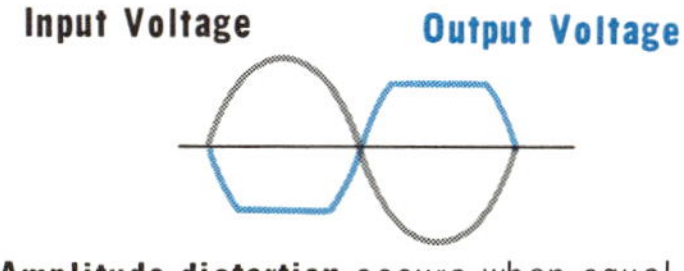

Amplitude distortion occurs when equal changes in the input signal do not produce proportionate changes in the output

For convenience of presentation, little difference in overall amplitude is shown. In actual amplifiers, the output voltages are many times greater than the inputs. The effects of the various types of distortion on the output waveforms, though, are as shown

summary

□ An amplifier circuit includes all of the circuit components that provide bias, as well as any used in compensating networks. □ According to use, the two main amplifier classifications are voltage amplifiers and power amplifiers. Voltage amplifiers increase the voltage level of a signal, while power amplifiers deliver large currents to their output load.

□ According to bias, amplifiers are classified as class A, class B, class AB, and class C. □ According to the coupling method used, amplifiers are classified as resistance capacitance (RC) coupled, impedance coupled, transformer coupled, and direct coupled. □ According to circuit configuration, amplifiers are classified on the basis of the principal elements returned to a-c ground. For example, grounded-grid amplifier. □ According to bandwidth, there are two principal types of amplifiers: those that are tuned and amplify a restricted range of frequencies; and those that are untuned and amplify a wide range of frequencies. □ According to frequency, amplifiers are classified as direct-current (d-c) amplifiers, audio-frequency (a-f) amplifiers, intermediate-frequency (i-f) amplifiers, radio-frequency (r-f) amplifiers, and video-frequency amplifiers.

□ The output of an ideal amplifier is identical to the input except for an increase in amplitude, and often there is a 180-degree phase-reversal. In practical amplifiers, though, some distortion always occurs. □ There are three main types of amplifier distortion: frequency distortion, phase distortion, and amplitude distortion. □ An extreme example of amplitude distortion is an input signal that drives an amplifier into saturation. □ Frequency distortion occurs when some frequency components of the signal are amplified more than others. □ Phase distortion occurs because coupling circuits often shift some frequency components of the signal more than others.

review questions

1. What is the difference between a voltage amplifier and a power amplifier?
2. Name four types of amplifiers classified by frequency.
3. How does saturation cause distortion?
4. What is the difference between a class AB and a class B amplifier?
5. Name the four basic coupling methods.
6. How do tuned and untuned amplifiers differ?
7. What are *compensating networks*?
8. What causes frequency and phase distortion in an amplifier?
9. Is distortion in an amplifier always undesirable?
10. Which is more efficient: a class A or a class C amplifier?

audio amplifier circuits

Audio amplifier circuits of some type are used in every piece of electronic equipment whose function is to transmit, receive, or otherwise process sound signals. Audio amplifiers are used in radio and television receivers, transmitter modulators, tape recorders, phonographs, and public address systems. In radio and television receivers, the audio amplifier circuits receive the audio signal after it has been removed from the AM or FM carrier by detection. They amplify this weak signal to the level required to drive the receiver loudspeaker. In tape recorders and phonographs, the audio amplifier circuits receive their input signal from *pickup devices,* which develop audio signal voltages that correspond to the audio intelligence contained on the tapes or records. The circuits amplify these audio signals to the point where they can suitably drive the equipment loudspeaker.

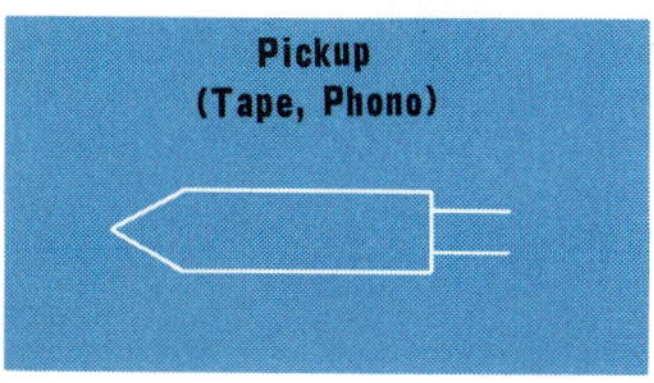

These are the schematic diagram symbols for the input and output devices commonly used with audio amplifiers

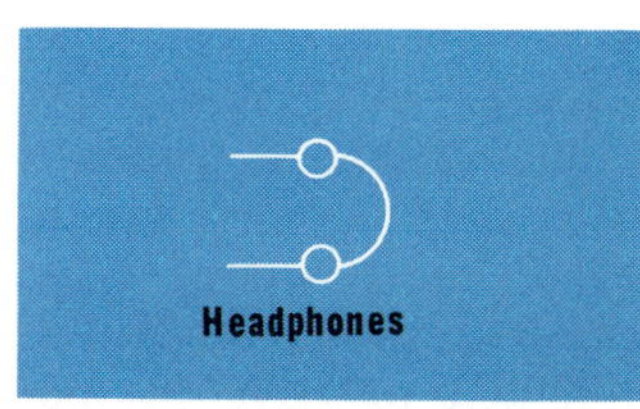

In other applications, such as transmitter modulators, the input to the audio amplifier circuits is often supplied by a microphone. The modulator circuits amplify the low-level output of the microphone so that when it modulates the r-f carrier, the desired percentage, or level, of modulation will be realized. The audio circuits of a public address system amplify the microphone output so that it can be sent over the system interconnecting wires and provide satisfactory driving power for the system loudspeakers.

Although all audio amplifiers distort the amplified signal to some extent, the degree of distortion that can be tolerated varies widely depending on the application. In the high-fidelity reproduction of music, for example, very little distortion is permissible. But in public address systems, where intelligibility rather than fidelity is important, a considerable amount of distortion can be tolerated.

the basic two-stage audio amplifier

Both a tube and a transistor version of a basic two-stage audio amplifier are shown on the following page. A functional diagram of the tube amplifier is shown below. As you can see from the functional diagram, the basic tube amplifier circuit consists of two stages: an audio *voltage* amplifier and an audio *power* amplifier. Both stages are biased for *class A* operation to minimize distortion of the amplified signal.

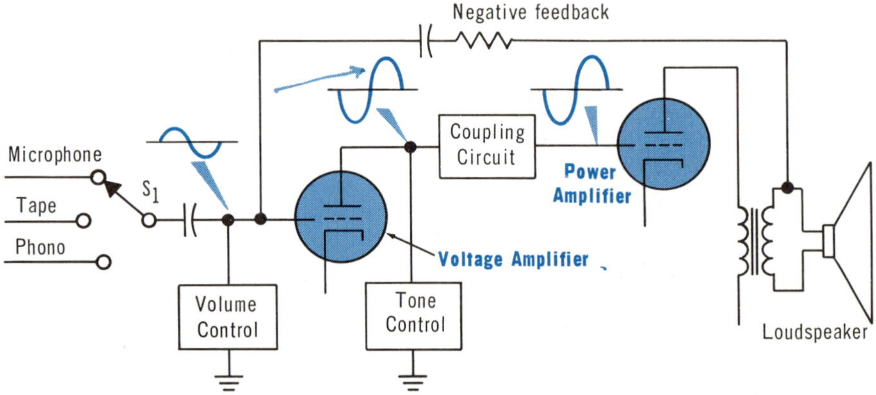

This functional diagram shows the path that the signal follows in the basic two-stage audio amplifier illustrated on the next page

The voltage amplifier stage receives an audio input signal from a microphone, a magnetic tape unit, or a phono pickup cartridge, depending on the position of switch S_1. The purpose of the voltage amplifier is to increase the *voltage level* of the input signal to the point where it is sufficient to drive the power amplifier stage. This voltage increase is necessary, since the power amplifier requires a relatively large input voltage in order to develop a large current or power output. The amplified signal voltage at the output of the voltage amplifier is coupled to the power amplifier, which produces the audio power required to drive the loudspeaker. *Transformer coupling* is used between the output of the power amplifier and the loudspeaker to obtain the proper impedance match, and, therefore, maximum transfer of power.

Additional features of the two-stage audio amplifier include a *tone control* circuit, *volume control,* and the use of *negative feedback* in the power amplifier stage. The two-stage amplifier shown here and on the next page illustrates many of the characteristics and elements of audio amplifier circuits in general. The various portions of the basic amplifier, together with their function and operation, are described in detail on the following pages. In addition, for each portion of the circuit, other methods of accomplishing the same function will be covered.

the basic two-stage audio amplifier (cont.)

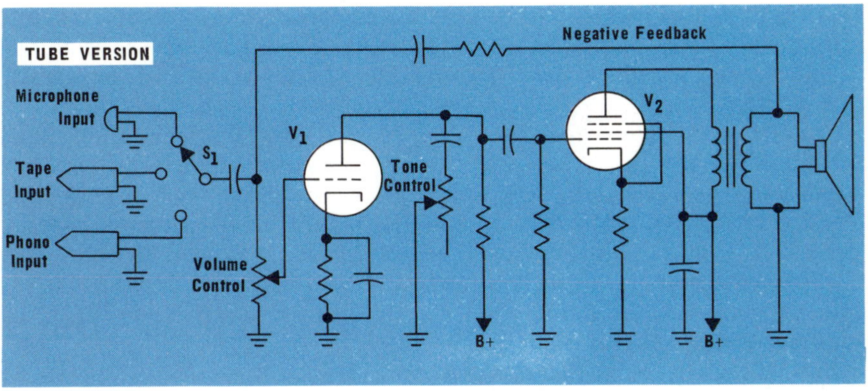

An important difference between two-stage tube and transistor amplifiers is the type of coupling used between stages

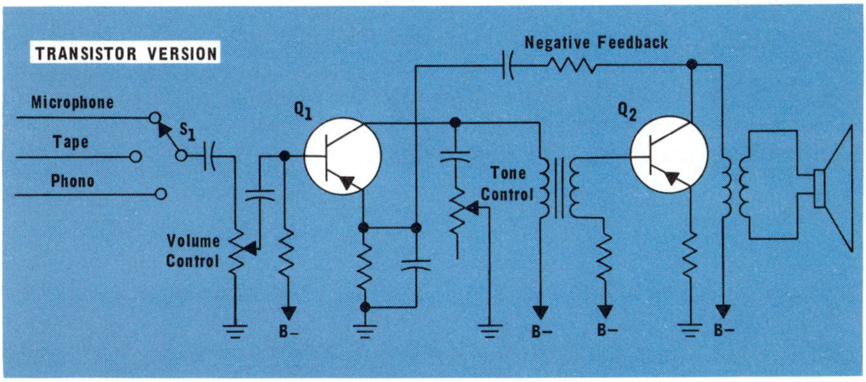

An important difference between the tube and transistor versions of this two-stage audio amplifier is the coupling used between the stages. Although RC coupling is satisfactory for the tube version, it is impractical when transistors are used because of the large difference in input and output impedance that is characteristic of transistor amplifiers. If RC coupling was used, the decrease in overall amplification caused by the impedance mismatch would make it necessary to use more than two amplifier stages to produce a gain equivalent to that of the two-stage tube amplifier. By using transformer coupling, though, the impedances can be matched to produce maximum amplifier gain.

Sometimes, the first transistor stage is called a driver, or a driver stage. The driver is biased to produce maximum *power gain,* while the power amplifier is biased to produce maximum *power output.*

the a-f voltage amplifier

Triodes are used for a-f voltage amplifiers in applications requiring relatively *low voltage gain.* When *high gain* is required, *pentodes,* with their large values of mu, are usually used. A typical pentode a-f voltage amplifier is shown in A. The pentode stage is similar to the triode shown on the previous page, with the exception of its screen and suppressor grid circuits.

The stage is biased for class A operation, which means that plate current flows for the *complete* cycle of the input signal and the grid is *never* driven positive by the input signal. The input voltage is applied across capacitor C_1 and volume control R_g, which together form a voltage divider. At audio frequencies, the reactance of C_1 is low, so that most of the signal voltage appears across R_g. Resistor R_g is variable and is adjusted to tap off a signal level for the desired loudness level at the amplifier loudspeaker.

The voltage between the movable arm of R_g and ground is applied to the grid of the stage. *Cathode bias* for the stage is provided by cathode resistor R_k and capacitor C_k. The variations in plate current produced by the input signal cause the voltage drop across load resistor R_L to have an a-c component. This a-c component is the amplified output of the stage and is coupled to the a-f power amplifier through capacitor C_2. C_2 passes the a-c component and blocks the dc. Resistor R_s is the screen dropping resistor and provides the proper screen grid operating voltage. Capacitor C_s bypasses the variations in screen current to ground, preventing them from causing variations in screen voltage.

A Pentode A-F Voltage Amplifier

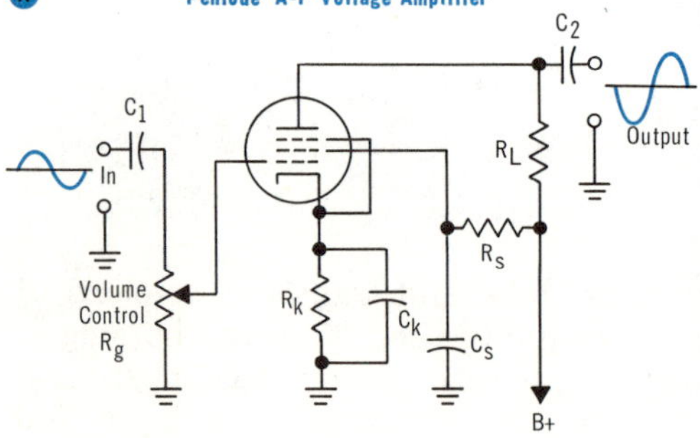

B Transistor A-F Amplifier

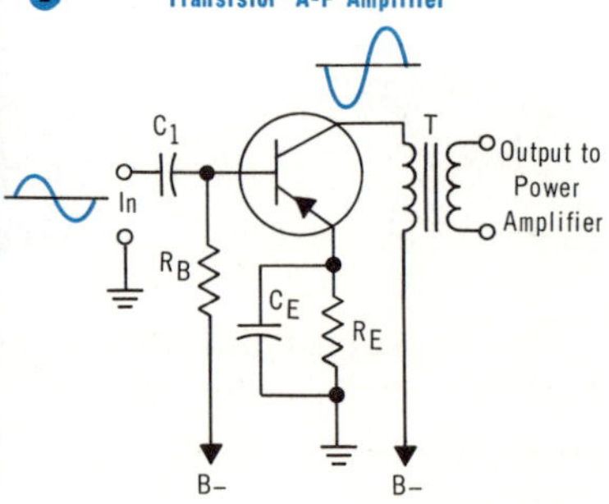

In the driver amplifier of the transistorized two-stage amplifier shown in B, the input signal is coupled through capacitor C_1 to the base-emitter circuit of the grounded-emitter amplifier. The amplified output signal is developed across the primary of transformer T in the collector circuit. The turns ratio of T is such that it matches the output impedance of the driver to the input impedance of the power amplifier. Resistor R_E and capacitor C_E in the emitter circuit of the stage maintain bias stability by providing negative d-c feedback to the base.

the a-f power amplifier

In the tube version of our two-stage audio amplifier, the power amplifier is a high-mu pentode biased for class A operation. This a *single-ended* power amplifier, since a *single tube* is used. You will learn later about other types of power amplifiers, such as *push-pull* amplifiers.

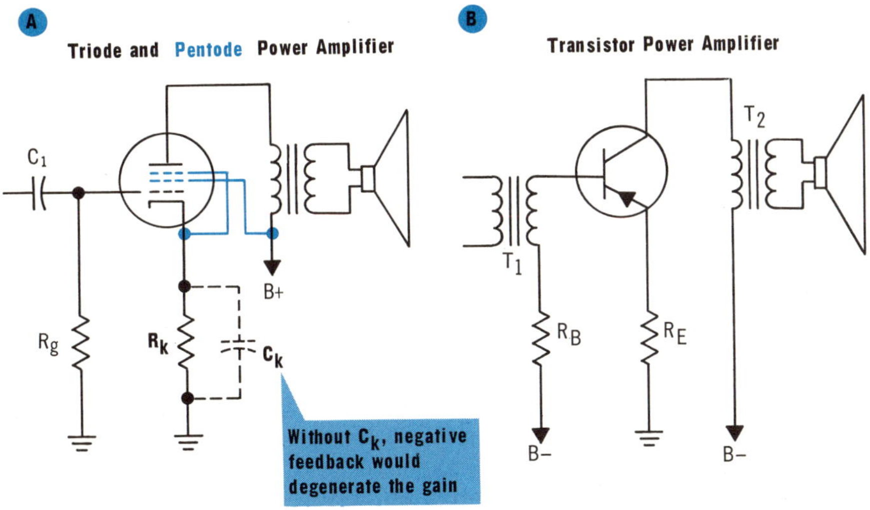

A single-ended pentode power amplifier is shown in A. If you ignore the elements shown in color, you can see how *triodes* can also be used for a-f power amplifiers. However, because of their low gain, triodes require larger input signal voltages. The input signal is coupled to the control grid by C_1 and R_g. Bias for the stage is provided by cathode resistor R_k and bypass capacitor C_k. Since *maximum power* is to be developed in the plate circuit, the plate load, which is the impedance offered to the signal current by the transformer primary, is relatively low. Transformer coupling of the output signal to the loudspeaker is necessary to match the amplifier output impedance to the very low impedance of the loudspeaker. A *step-down* transformer is used, so a relatively low-voltage, high-current signal is applied to the loudspeaker.

The a-f power transistor differs from the a-f driver mainly in size. The power transistor is larger, since it must handle considerably more power. In the basic amplifier, shown in B, the power-stage input is applied to the base-emitter circuit by transformer T_1. The amplified output is coupled to the loudspeaker by impedance-matching transformer T_2. Both a-c and d-c negative feedback are provided by unbypassed emitter resistor R_E. The d-c feedback provides bias stability; the a-c feedback reduces distortion. This is covered in more detail later in this volume.

the speaker

The output of our basic audio amplifier drives a loudspeaker, which we will refer to from now on as a *speaker*. The function of the speaker is to convert the *current variations* delivered to it by the power amplifier into *sound waves*.

As you know, sound waves are made up of alternate compressions and rarefactions of the air, which cause the sensation of sound when they enter the ear. In a speaker, the sound waves are produced by a paper *cone*, or horn, that moves back and forth in accordance with the signal variations. When the cone moves forward, it pushes the air in front of it, causing compressions; when the cone moves backward, it releases the pressure on the air in front of it, causing rarefactions.

Most speakers used today are of the *dynamic* type and are based on the interaction of two magnetic fields whose directions are at right angles to each other. There are two principal types of dynamic speakers: the *permanent-magnet* (PM) type and the *electrodynamic* type.

In the PM dynamic speaker, a strong magnetic field is produced by a *permanent magnet*. The flux lines of this field are concentrated in a soft iron core shaped as shown. Mounted in the gap between the pole pieces of the iron core is a coil, called the *voice coil*, which carries the audio signal current from the output of the a-f power amplifier. The voice coil is mounted in such a way that it is free to move back and forth for a small distance in a longitudinal direction. The speaker cone is attached to the voice coil and moves with it. The magnetic field produced by the signal current in the voice coil is at right angles to the field of the permanent magnet. The two fields attract or repel each other, depending on the amplitude and polarity of the signal current. This attraction or repulsion causes an inward or outward movement of the voice coil and the cone, and thus produces sound waves that correspond to the variations in signal current.

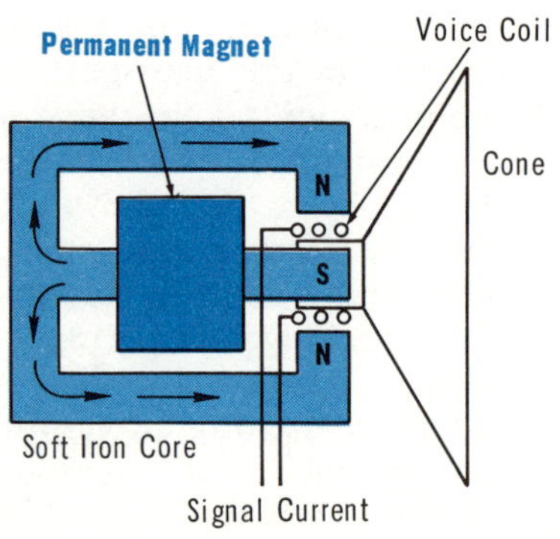

PM SPEAKER

The signal current through the voice coil produces its own varying magnetic field, which interacts with the main field to cause the cone to vibrate and produce sound waves

the speaker (cont.)

The voice coil consists of only a few turns of low-resistance wire and has a relatively *low impedance* throughout the audio frequency range. At the lower audio frequencies, the impedance is generally somewhere between 2 and 8 ohms. It is because of this low impedance of the voice coil that an impedance-matching transformer must be used between the output of the power amplifier and the speaker.

A

ELECTRODYNAMIC SPEAKER

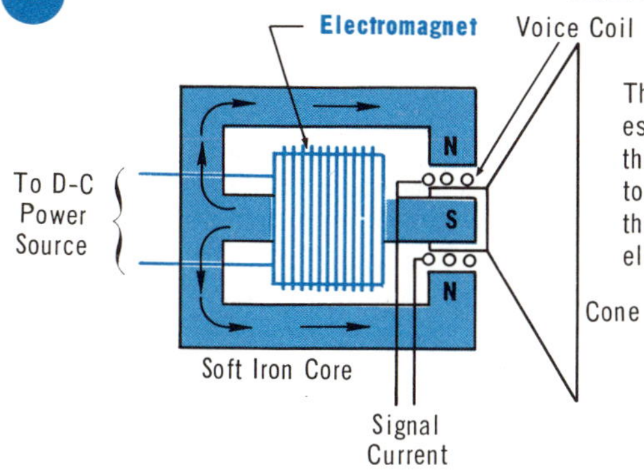

The PM and the electrodynamic speaker are essentially similar. The basic difference is that the PM speaker uses a permanent magnet to produce the constant magnetic field, while the electrodynamic speaker uses an electromagnet

The electrodynamic speaker is similar to the PM speaker. The basic difference is that an *electromagnet* is used to generate the constant magnetic field instead of a permanent magnet. This, of course, requires that a d-c power source supply energizing current to the electromagnet. An electrodynamic speaker is shown in A. It is not used as often as a PM speaker because it is more expensive.

B

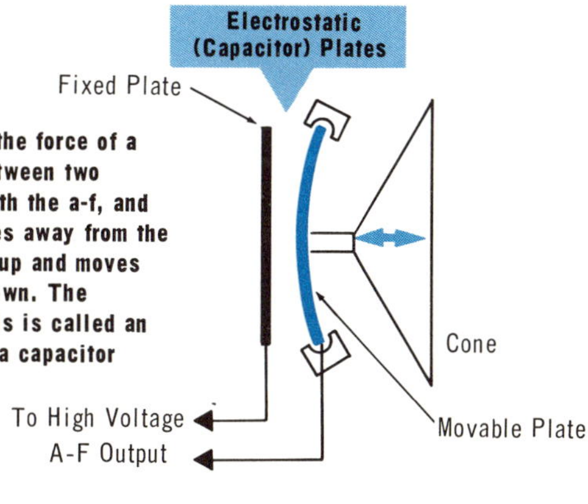

Another type of speaker relies on the force of a high-voltage electrostatic field between two plates. The high voltage varies with the a-f, and the spring-like movable plate moves away from the fixed plate when the voltage goes up and moves toward it when the voltage goes down. The movable plate drives the cone. This is called an ELECTROSTATIC SPEAKER, or a capacitor speaker

the microphone

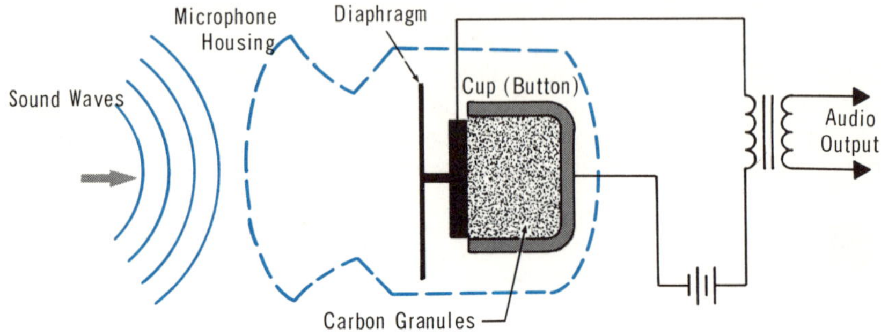

The carbon granules of a carbon microphone make up a resistance whose value is varied by the pressure of the sound waves

The principal disadvantage of the carbon microphone is that it requires an external voltage source

One of the inputs to our basic audio amplifier is from a microphone, which converts sound waves into corresponding electrical signals. There are many types of microphones in use today. The type chosen for any particular application depends on mechanical characteristics, such as size, weight, and ruggedness, as well as electrical characteristics, such as frequency response, impedance, and sensitivity.

A microphone's frequency response is the range of audio frequencies over which the microphone can respond uniformly. The impedance of a microphone is important with respect to the input impedance of the circuit that the microphone is driving. To deliver maximum output to its load, these impedances must be of the same order. Otherwise, impedance-matching devices, such as transformers, are required. The sensitivity of a microphone is a measure of how much electrical power the microphone delivers to a load having a matched impedance for a given input sound level. High sensitivity thus means a high electrical output for a given sound level.

Probably the most common type of microphone in use is the *carbon microphone,* which operates on the principle that the *electrical resistance* of carbon granules varies as the *pressure* on the granules varies. As shown, a carbon microphone consists of a small cup of carbon granules and a diaphragm that exerts varying pressure on the granules when it is struck by sound waves. To produce an output voltage, the cup is connected in a *series circuit* with a battery and the primary of a transformer. When no sound waves strike the diaphragm, a steady d-c current flows in the circuit, so there is no audio output in the transformer secondary. But sound waves striking the diaphragm vary the resistance of the carbon granules and a varying d-c current flows in the circuit. The *variations,* coupled to the transformer secondary, are the microphone output. This type of carbon microphone is called a *single-button* microphone, since it contains a single cup, or button, of carbon granules.

other types of microphones

Other frequently used types of microphones are the *dynamic microphone* and the *crystal microphone*. As shown in A, the dynamic microphone is similar to the PM speaker, except that it converts sound energy into electrical energy. The diaphragm of the microphone, which corresponds to the cone of the PM speaker, moves, or vibrates, when struck by sound waves. The voice coil is attached to the diaphragm and moves with it. Since the voice coil is positioned in the magnetic field of a permanent magnet, it cuts flux lines when it moves. As a result, a voltage is induced in the voice coil corresponding to the sound waves striking the diaphragm.

The crystal microphone, shown in B, operates on the principle that certain crystal materials generate a voltage when *mechanical stress* is applied to them. This is called the *piezoelectric effect*. A basic crystal microphone consists of a crystal, a diaphragm that contacts one side of the crystal, and an electrode plate that contacts the other side of the crystal. Sound waves striking the diaphragm cause the diaphragm to apply a varying stress to the crystal. The crystal, in turn, generates a voltage that has essentially the same waveform as the sound waves.

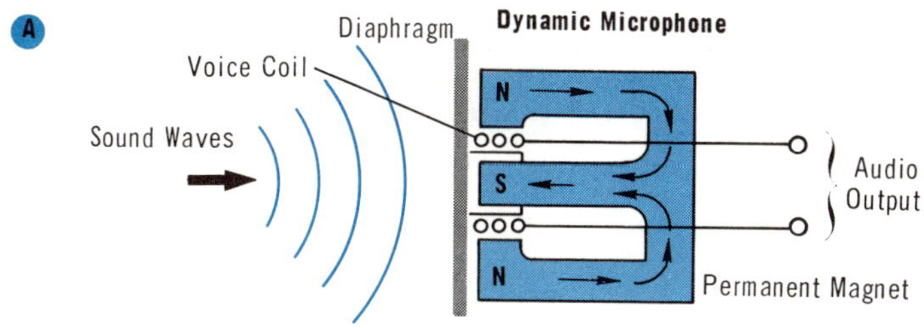

In the dynamic microphone, the sound waves cause the voice coil to move in the field of the permanent magnet. This induces a voltage in the voice coil that corresponds to the sound wave variations

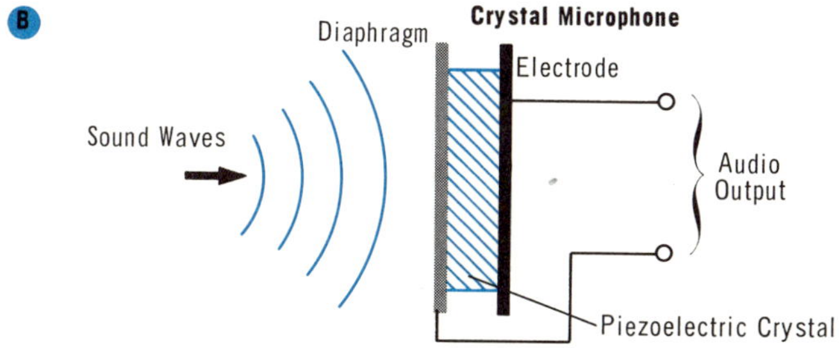

In crystal microphones, the sound waves cause mechanical stress of a piezoelectric crystal. The crystal produces a voltage that corresponds to the sound wave variations

the phono and magnetic-tape pickups

Our basic two-stage audio amplifier is equipped with both a phono and a magnetic-tape input. Both of these inputs, like the microphone input, consist of signal voltages that vary at an audio rate in accordance with the signal intelligence. The only difference between any of the inputs is the manner in which they are generated before being applied to the audio amplifier.

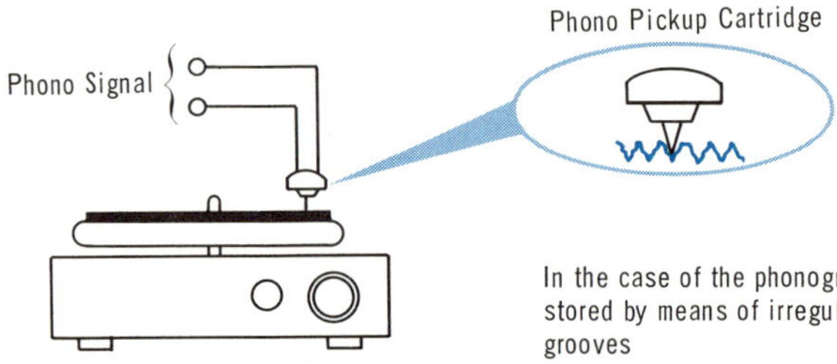

In the case of the phonograph, the signals are stored by means of irregularities in the record grooves

Both phono and magnetic-tape pickups generate electronic signals that have been "stored" by nonelectrical means

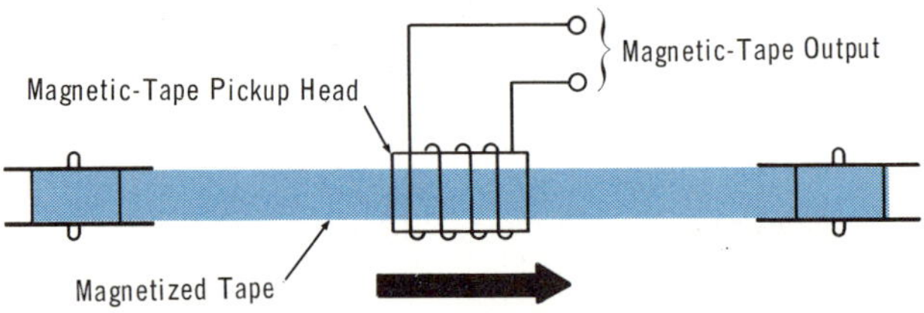

On the magnetic tape, the signals are stored by magnetized particles

The phono input is generated by a pickup cartridge that converts the *movements* of the phono needle in the record grooves into a corresponding, varying voltage. This signal voltage may be on the order of millivolts when magnetic pickup cartridges are used, and nearly one volt when crystal cartridges are used.

The magnetic-tape input is generated by a pickup head in a tape recorder. Essentially, a pickup head is a coil that has a voltage *induced* in it when a magnetized tape moves past it. The magnitude and polarity of the induced voltage correspond to the variations in tape magnetization.

volume control

The volume control of both versions of our two-stage amplifier consists of a *variable resistor* that is manually adjusted to provide the desired listening output level from the speaker. The volume control operates by setting the *amplitude* of the signal input to the first stage of the amplifier. The signal current flows through the volume control resistor and develops a voltage across it. The larger the resistance, the greater is the voltage drop, and the larger is the input signal to the stage.

The basic difference between the volume control circuits used in our tube and transistor circuits is the manner in which the signal voltage picked off the volume control resistor is applied to the first amplifier stage. In the tube circuit, the signal voltage can be applied *directly* to the grid. In the transistor circuit, the signal is *capacitively coupled* to the base circuit. If this capacitive coupling were not used, the base-emitter bias of the transistor would change every time the volume control was adjusted. The capacitor prevents this from happening by isolating the volume control resistor from the d-c bias circuit of the transistor.

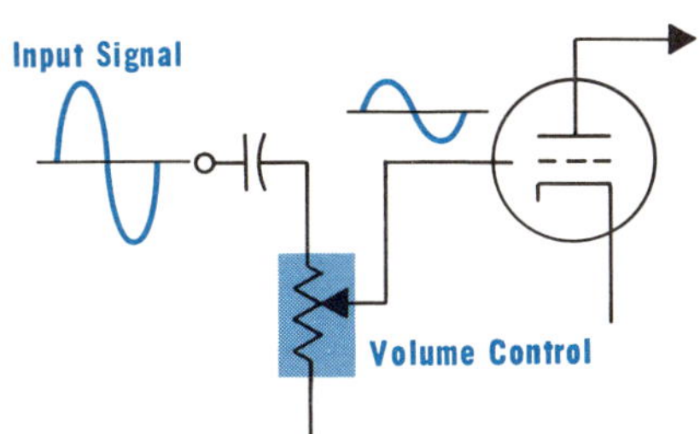

In this tube circuit, a portion of the input signal voltage is applied directly to the grid

The volume control is usually a variable resistor that applies a portion of the input signal to the first audio amplifier

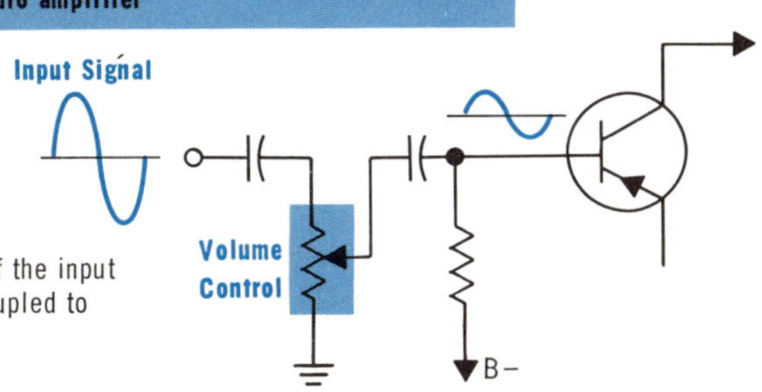

In transistor circuits, a portion of the input signal voltage is capacitively coupled to the base circuit

In tube circuits, the a-f signal taken off the volume control can also be capacitively coupled. There is little difference in most tube circuits whether it is or is not. In tube amplifier stages that use grid-leak or contact bias, capacitive coupling should be used or the tube will be affected by the setting of the volume control in the same fashion that a transistor is.

tone control

Tone controls are essentially *variable filter circuits* that alter the output frequency response of an audio amplifier. A tone control may be included in an audio amplifier to permit the output tonal quality to be adjusted for most pleasant listening, or it may be included out of necessity. An example of this would be in an equipment that uses a small-diameter speaker. Such speakers accent the high frequencies of the audio output and must generally be compensated for by tone controls that attenuate or reduce the high frequencies. Strictly speaking, these circuits are really filter circuits instead of tone controls, since they are not variable. The basic principles on which they operate, however, are the same as those involved in tone controls.

The simplest type of tone control consists of a capacitor and a variable resistor connected between the plate of a tube, or the collector of a transistor, and ground. This is shown in A. Such a tone control is a *treble control,* since it acts on the *high frequencies* of the audio signal. The capacitor presents a relatively low reactance to the higher frequencies and a high reactance to the lower frequencies. The setting of the variable resistor then determines to what degree the higher frequencies are shunted to ground. Very little of the low-frequency energy is shunted to ground, regardless of the setting of the variable resistor, because of the high reactance the capacitor offers to these frequencies.

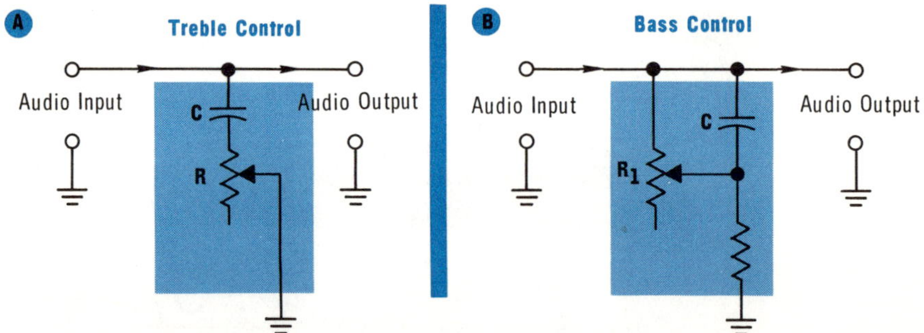

Tone controls are essentially variable filter circuits that present different impedances to different frequencies

A simple *bass control* is shown in B. This type of tone control acts at the *lower end* of the frequency range. The value of capacitor C is such that it has a large reactance at the lower audio frequencies and a low reactance at the higher frequencies. This causes the voltage drop between the top of C and ground to be larger at low frequencies than at high frequencies. Variable resistor R_1 shunts capacitor C, and essentially reduces the boosting effect of C on the low frequencies. The setting of R_1 determines just how much the effect of C is reduced.

More elaborate tone controls combine *both* treble and bass control circuits.

negative current feedback

The feedback voltage decreases both the gain and the distortion of the power amplifier stage

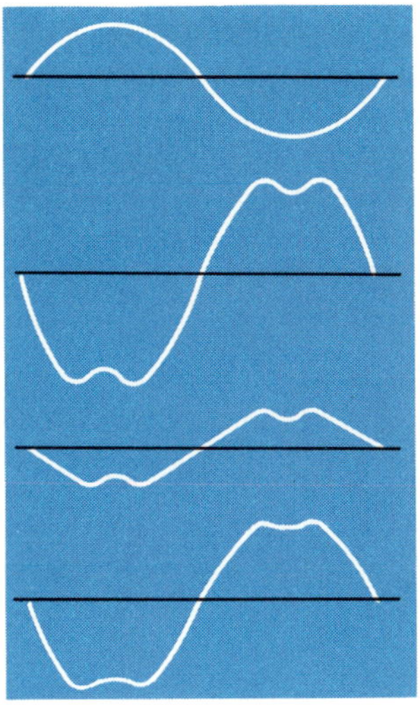

Input Signal

Output Signal If Negative Feedback Were Not Used

Feedback Signal When Negative Feedback Is Used

Output Signal When Negative Feedback Is Used

As you learned, negative feedback in an amplifier is a process whereby a portion of the output signal is fed back to the input of the amplifier *180 degrees out of phase* with the input signal. Because of the phase difference, the feedback signal *cancels* part of the input signal, so the gain of the stage is less than it would be if feedback were not used. However, as shown, the feedback not only cancels a portion of the input signal, but it also tends to *cancel distortion* in the output signal caused by nonlinearities in the stage. As a result, negative feedback is often used in audio amplifiers in spite of the decrease in gain that it causes.

Negative feedback in a grounded-emitter amplifier is accomplished by using an unbypassed emitter resistor. The audio signal variations that appear across the emitter resistor produce a feedback voltage that opposes the input signal

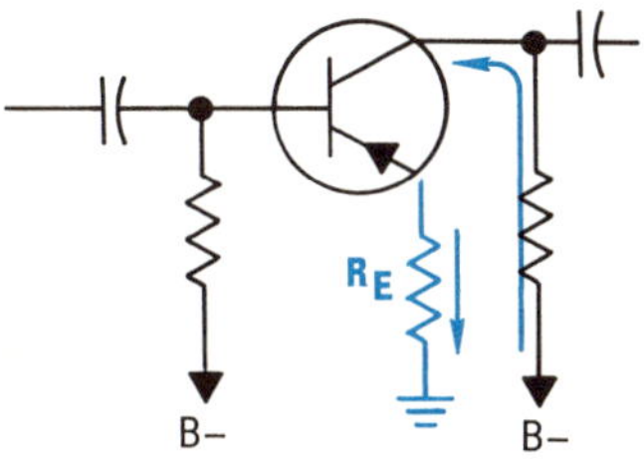

Negative feedback cannot be produced in this way in a grounded-base or grounded-collector amplifier. In these types of amplifiers, an unbypassed resistor in series with the common element would produce a feedback voltage that aided the input signal. The feedback would therefore be positive

In the transistor version of our basic two-stage audio amplifier, negative feedback is used in the power amplifier stage. One kind of feedback is produced by using an *unbypassed emitter resistor*. You recall that the emitter bypass capacitor charges and discharges with the audio variations of the emitter current. This, in effect, bypasses the audio variations around the emitter resistor and maintains a steady current and, therefore, voltage drop, in the emitter circuit.

negative current feedback (cont.)

If the emitter resistor is unbypassed, though, the current in the emitter circuit, as well as the voltage drop across the emitter resistor, *varies* according to the audio signal. These audio variations in the emitter circuit are 180 degrees out of phase with the input signal applied to the base, and oppose or cancel a portion of the input signal. This cancellation, or *degeneration,* as it is called, significantly reduces distortion of the signal introduced within the transistor.

With current feedback, the audio variations in tube current produce a varying voltage drop across the unbypassed cathode resistor. This voltage includes the signal variations plus the nonlinear distortion introduced in the tube. Since the cathode voltage is 180° out of phase with the signal voltage on the grid, the feedback voltage opposes the grid voltage. In this way the feedback voltage decreases both the gain and the distortion of the stage

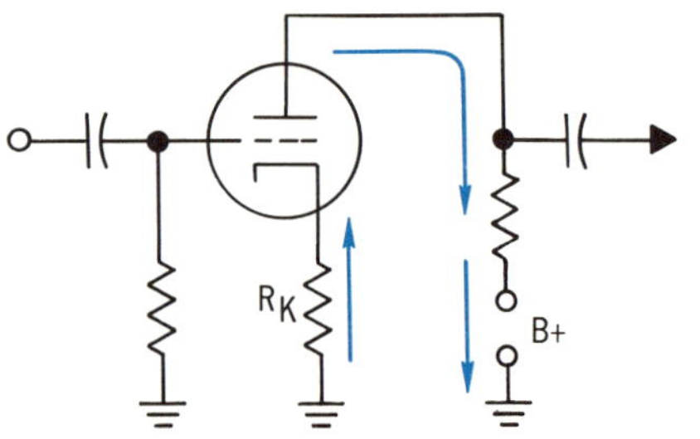

Negative feedback produced by an unbypassed emitter or cathode resistor is called *current feedback,* since it is derived from the signal current variations in the stage. As you will see, negative feedback can also be produced by transferring a portion of the output voltage of a stage directly back to the input. This is called *voltage feedback.*

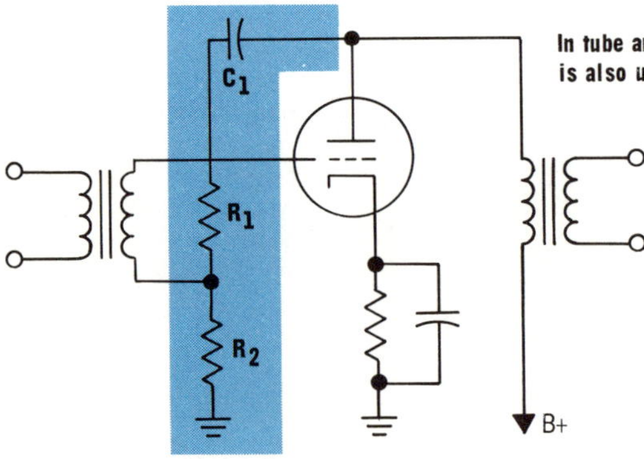

In tube amplifiers, negative voltage feedback is also used

In this circuit, a portion of the output voltage is fed back to the grid circuit by C_1, R_1, and R_2. C_1 passes only the audio signal variations, thereby isolating the grid circuit from the plate B+ supply. R_1 and R_2 form a voltage divider. Their values are chosen so that the desired feedback voltage is developed across R_2, which is in series with the secondary of the input transformer as far as the input signal is concerned. Because of the 180° phase shift introduced by the tube, the feedback voltage across R_2 is in phase opposition to the input signal voltage developed across the secondary of the input transformer. The feedback voltage is therefore negative

negative voltage feedback

Negative feedback in an amplifier can be accomplished not only by using the current variations in the stage to transfer part of the output signal back to the input, but also by feeding a portion of the *output voltage* directly back to the input. A frequently used grounded-emitter circuit employing negative voltage feedback is shown.

In this circuit, feedback resistor R_F is connected between the collector and the base input circuit. Since resistor R_F is not bypassed by a capacitor, it provides a-c, or signal, feedback, as well as d-c feedback for bias stability. When a *positive-going* input signal is applied to the base, the output voltage at the collector becomes more *negative* as a result of the 180-degree phase shift introduced by the stage. A portion of this output signal voltage is fed back to the base by the feedback resistor, and therefore *opposes* the input signal, causing degeneration.

Feedback resistor R_F transfers a portion of the output signal voltage back to the amplifier input

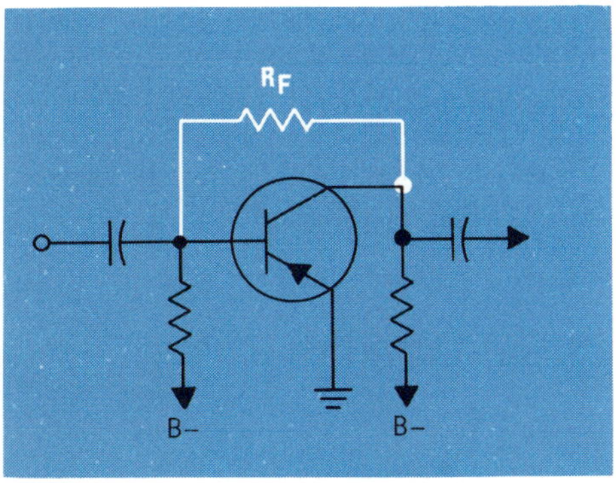

The feedback voltage is out of phase with the input signal because of the 180° phase shift introduced by a grounded-emitter amplifier

The situation is similar when a negative-going input signal is applied to the base. A portion of the positive increase in collector voltage is coupled back to the base by the feedback resistor. Again, you should be aware that this type of voltage feedback would cause regeneration, or positive feedback, in grounded-base or grounded-collector amplifiers, unless some means is taken to shift the phase of the feedback voltage. This could be done, effectively, by using the feedback signal from one stage to degenerate the input of another stage. Quite often, the output of a string of amplifiers is fed back to one of the input stages to obtain good negative feedback.

summary

☐ A basic two-stage audio amplifier consists of an audio voltage amplifier and an audio power amplifier. Both stages are biased for class A operation to minimize distortion. ☐ RC coupling is generally used in tube-type audio amplifiers. In transistor amplifiers, an interstage transformer is generally used. ☐ The first stage of a transistor audio amplifier is called a driver, and is biased to produce maximum power gain. The power amplifier is biased to produce maximum power output. ☐ In tube-type audio amplifiers, triodes are used for the voltage amplifier in applications requiring relatively low voltage gain, while pentodes are generally used when high gain is required. The power amplifier is usually always a pentode.

☐ The speaker converts the current variations produced by the power amplifier into sound waves. ☐ There are two principal types of dynamic speakers: the permanent-magnet (PM) type and the electrodynamic type, which uses an electromagnet. ☐ The microphone converts sound waves into corresponding electrical signals. ☐ The most common type of microphone in use is the carbon microphone. Other frequently used types of microphone are the dynamic microphone and the crystal microphone, which operates on the principle of the piezoelectric effect. ☐ The phono input is generated by a pickup cartridge, which converts movements of the phono needle in the record grooves into a varying voltage. Either a magnetic pickup or crystal pickup can be used. ☐ The magnetic-tape input is generated by a pickup head in a tape recorder.

☐ The volume control on an audio amplifier is used to control the output level from the speaker. It does this by varying the input signal applied to the first stage. ☐ The tone control alters the output frequency response to permit the tonal quality of the audio amplifier to be adjusted. ☐ Negative feedback significantly reduces distortion in an audio amplifier. It does this by canceling distortion introduced by the amplifier.

review questions

1. How does the carbon microphone operate?
2. What is meant by the *sensitivity* of a microphone?
3. On what principle is the crystal microphone based?
4. What is meant by a *pickup cartridge*? Name two types.
5. What is *negative feedback*? Why is it used?
6. What is a *PM speaker*?
7. In a two-stage transistor audio amplifier, what type of coupling is generally used between stages? In a tube amplifier?
8. How does the tone control work?
9. What is an electrostatic speaker?
10. What is a driver?

the push-pull power amplifier

In a conventional tube-type audio power amplifier, the input signal is applied to the control grid of the amplifier stage, and the amplified output is developed across the secondary of a transformer in the plate circuit of the stage. Such a circuit is often called a *single-ended amplifier* to distinguish it from a special type of amplifier called a *push-pull amplifier*. In a push-pull amplifier, the signal to be amplified is applied simultaneously to *two tubes,* and the amplified output signal is developed across a transformer that is connected *between* the plates of the two tubes.

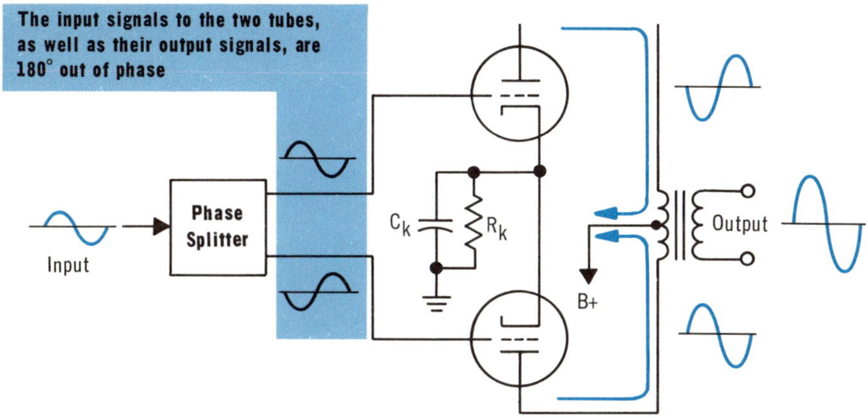

In a push-pull amplifier, the output signals of the two tubes combine across the common plate circuit transformer to produce a common signal that has a large power content and relatively small distortion

The basic principle on which the push-pull amplifier works is that the input signal is converted, before amplification, to *two* separate signals, which are identical except for a *180-degree phase difference.* Each of these signals is applied as the input to one of the tubes of the push-pull amplifier. Therefore, since the inputs to the two tubes are 180 degrees out of phase, so, also, are the outputs of the tubes. The output transformer in the plate circuits of the tubes is connected in such a way, though, that the audio components of the two separate plate currents combine to produce an *overall* audio signal having the *same waveshape* as the original input signal. The magnitude of this combined output signal is larger and contains less distortion than could be obtained with a comparable single-ended amplifier.

Push-pull amplifiers are normally operated class A, AB, or B. In class B operation, current for each tube flows for approximately *one half* of the input signal cycle. Each stage, though, amplifies a different half, so that the entire signal is reproduced in the output.

developing the output signal

The operation of transistor push-pull amplifiers is basically the same as the operation of tube types. In the class A grounded-emitter circuit shown, input signals 180 degrees out of phase are applied to the bases of transistors Q_1 and Q_2, respectively. The output signals from the two transistors are also 180 degrees out of phase and combine across the output transformer to produce an overall signal that is coupled by transformer action to the transformer secondary.

The way in which the outputs of the two transistors produce the overall output signal is shown in A through C. When analyzing these figures, remember that it is the *change* in the collector currents of the two transistors that creates a voltage across the transformer primary. These induced voltages combine to produce the output signal.

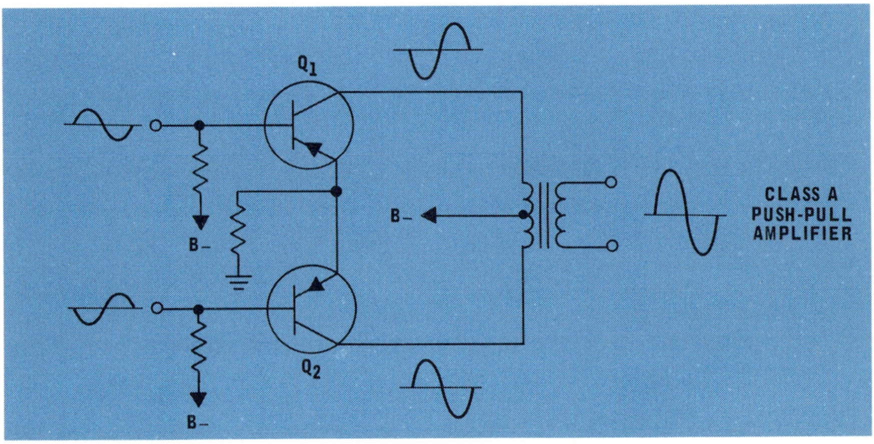

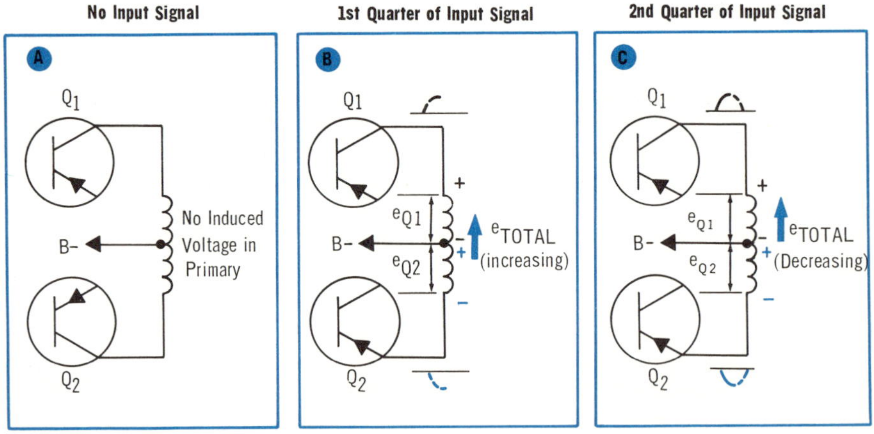

developing the output signal (cont.)

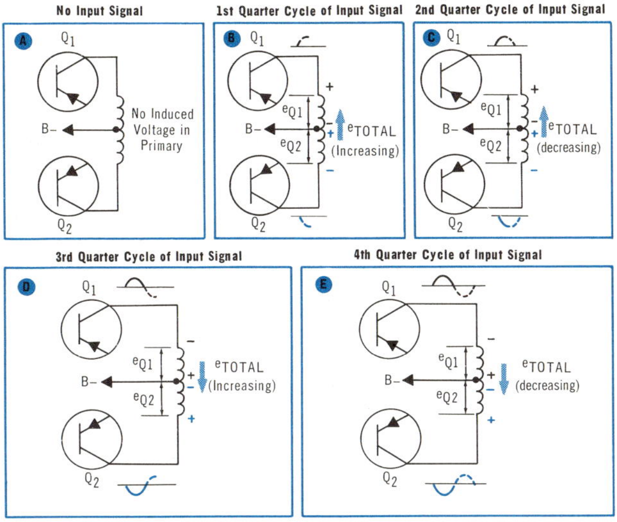

In A, there is *no* input signal, so d-c current flows from the B– supply through one half of the primary to the collector of Q_1 and through the other half of the primary to the collector of Q_2. This d-c current induces no voltage in the transformer primary. Actually, no magnetic field at all is built up around the primary, since the magnetizing forces in the two halves of the primary effectively *cancel* each other.

During the first quarter cycle of the input signal, as shown in B, collector current increases in Q_1 and decreases in Q_2. These changes in current cause induced voltages in the two halves of the primary. These voltages have polarities as shown. You can see from the polarities that the induced voltages add to produce one voltage across the primary. The amplitude of the overall voltage increases at the same rate as the current varies.

In the second quarter cycle of the input signal, shown in C, collector current decreases in Q_1 and increases in Q_2. These current changes induce voltages in the primary that have the same polarity as during the first quarter cycle, but which are decreasing in amplitude. The induced voltages have like polarities with respect to the primary winding, so they add together to produce the overall output voltage.

During the second half cycle of the input signal, as shown in D and E, the conditions that exist in the transformer primary are opposite to those existing during the first half cycle. The two induced voltages again always have like polarities, although the polarity is opposite to that of the first half cycle.

the cathode-coupled push-pull amplifier

A type of push-pull amplifier that does not require two separate out-of-phase inputs is shown. This circuit uses a grounded-cathode stage that drives a grounded-grid stage by means of a *common-cathode resistor*. The single input signal is applied to the grounded-cathode stage, V_1, by capacitor C_1 and resistor R_1. As in a conventional grounded-cathode amplifier, this causes an *inverted* version of the input signal to be developed at the plate of V_1.

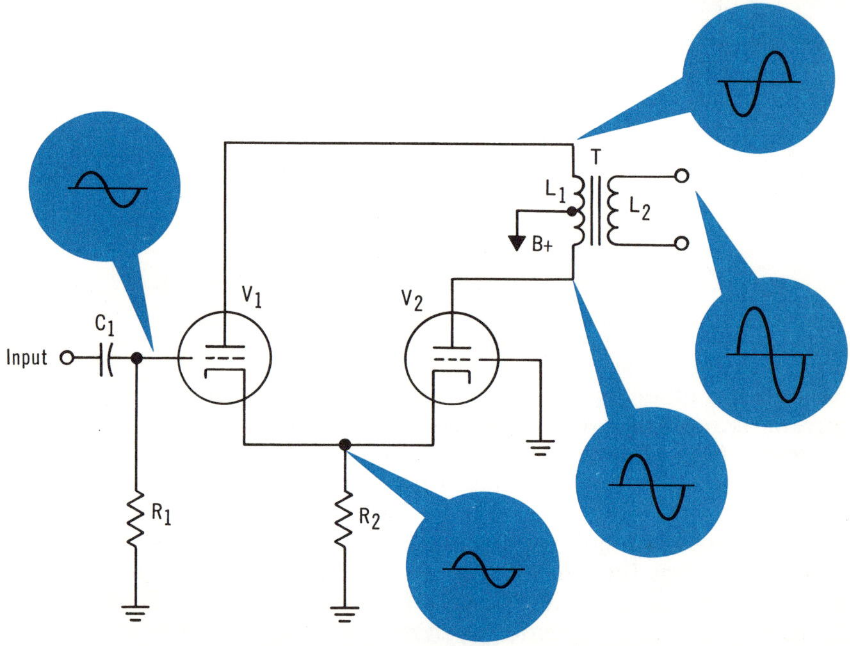

By using two different circuit configurations, one that produces a polarity reversal and another that does not, a push-pull amplifier can be operated without the need for a phase splitter

The variations in tube current through V_1 also produce a varying voltage across common-cathode resistor R_2 that is in phase with the input signal. Since the grid of V_2 is grounded, the voltage variations across R_2 serve as the input signal to V_2. As a result, the output signal at the plate of V_2 is *in phase* with the variations in cathode voltage, which are in phase with the input signal to V_1. Thus, the voltage variations at the plate of V_2 are in phase with the input signal, while those at the plate of V_1 are 180 degrees out of phase with the input signal. These voltage variations combine across transformer primary L_1, as described on the previous page.

the emitter-coupled push-pull amplifier

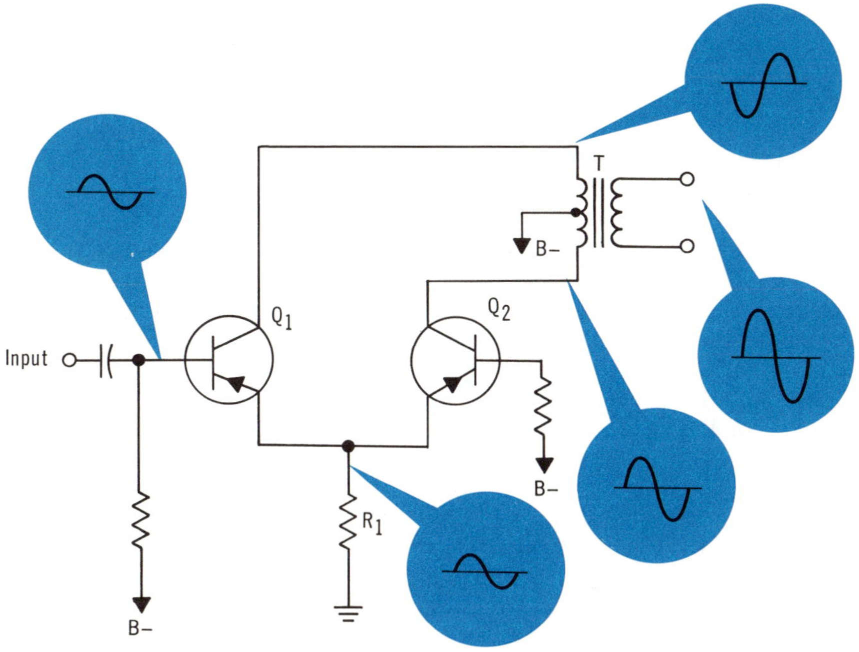

In both versions of this circuit, component values and characteristics must produce equal output voltages for each stage for a given input signal

A basic transistor circuit similar to the tube circuit shown on the preceding page is shown here. A common-emitter and a common-base stage are used, with the signal coupled from Q_1 to Q_2 by resistor R_1, which is in the emitter circuit of both Q_1 and Q_2. The collector signals of the two stages are 180 degrees out of phase, since a signal inversion takes place in a common-emitter stage but not in a common-base stage. The two out-of-phase signals combine across the primary of the output transformer to produce the overall output signal.

In both the tube and transistor versions of this circuit, the characteristics and values of the components must be such that equal output voltages are developed by both stages for a given input signal.

Frequently, a circuit of the type described above, instead of being used as a push-pull amplifier, is used to produce the out-of-phase input signals for a push-pull amplifier. The operation of the circuit is the same, except that the two outputs are usually capacitively coupled to the push-pull stages. Such a circuit is called a cathode-coupled or emitter-coupled phase splitter.

phase splitters

As was pointed out previously, phase splitters are the circuits that provide the separate inputs for push-pull amplifiers. Basically, a phase splitter receives as its input the signal to be amplified. It then delivers as its output *two* separate signals. One of the outputs is the *original* signal and the other is an *inverted* duplicate of the original signal. In other words, if the signal to be amplified is a positive-going square wave, the phase inverter delivers as its outputs both a positive-going square wave and a negative-going square wave.

In the case of a sine wave, the two outputs of a phase splitter can be considered as being 180 degrees out of phase with each other, since inverting a sine wave in polarity has the *same effect* as shifting it 180 degrees in phase. Because of this, it is common practice to speak of the outputs of a phase inverter as being 180 degrees out of phase, even though the inverted signal is the result of a polarity inversion rather than an actual phase shift.

As you will learn, phase splitting can be accomplished by transformers, or by tube or transistor amplifier circuits. When tube or transistor amplifier circuits are used, the circuits are also often called *paraphase amplifiers*. In addition, you will also often encounter what are called *inverter circuits*. Unlike a phase splitter, which has two outputs, an inverter delivers only a *single* output, which is an inverted version of its input signal. As you know, many tube and transistor amplifiers automatically invert a signal during the amplifying process. In an inverter circuit, the *inversion* of the signal is the basic circuit function, and any amplification that occurs is of secondary importance. In a conventional amplifier, the opposite is true.

A Phase Splitter . . .

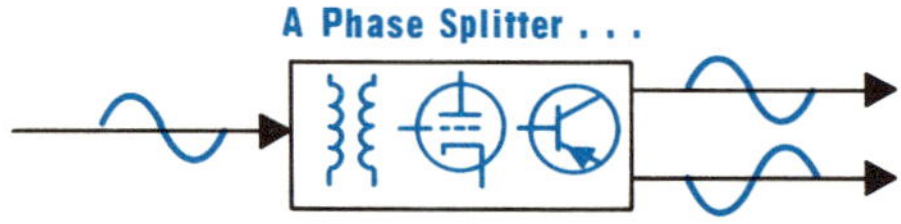

. . . is any transformer, tube, or transistor circuit that receives a single input and delivers two outputs of opposite instantaneous polarity

A Paraphase Amplifier . . .

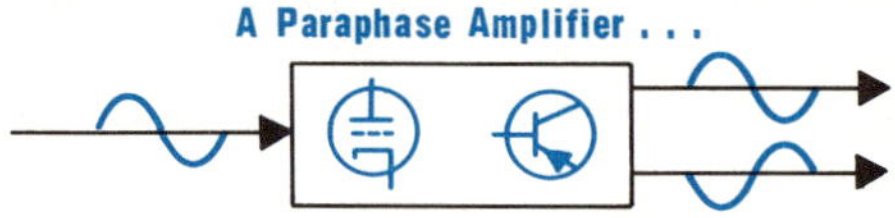

. . . is any tube or transistor amplifier circuit used as a phase splitter

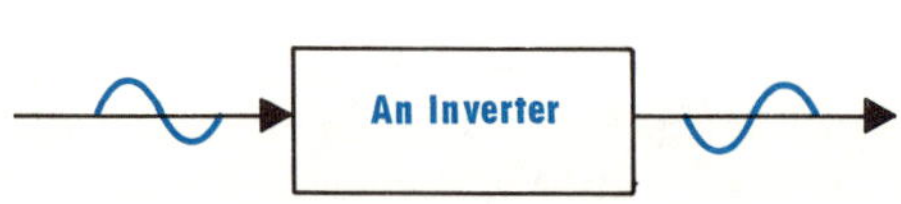

. . . is any circuit whose primary purpose is to deliver a single output of opposite instantaneous polarity to the input signal

transformer-type phase splitters

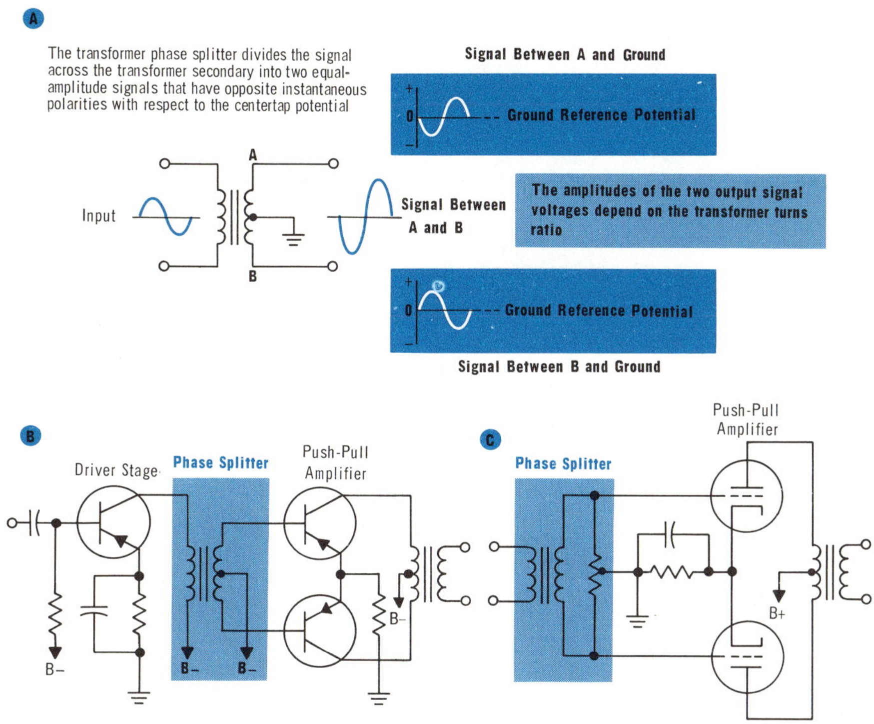

In a transformer-type phase splitter, a transformer having a *center-tapped secondary* is generally used. As shown in A, *two* separate signal voltages can be obtained from the secondary, since the secondary is centertapped to the zero potential reference point, or ground. One of these signals is taken from between the top of the secondary (point A) and the centertap, and the other is taken from between the bottom of the secondary (point B) and the centertap. Because of normal transformer action, points A and B always have opposite polarities.

The centertap is located at the *electrical center* of the secondary, so both of the output signals always have the same instantaneous amplitudes, as well as *opposite* instantaneous *polarities*. When these two signals are applied to the inputs of a push-pull amplifier (shown in B), they have the required characteristics for push-pull operation.

As shown in C, instead of directly centertapping the transformer secondary, a centertapped resistor connected across the secondary can be used to obtain the two output signals. The resistor then serves as the load for the signal induced in the transformer secondary, and the two opposite voltages across the balanced resistor are used as the outputs.

the single-tube balanced-type phase splitter

In a single-tube, balanced-type phase splitter, a single stage, often called a single-tube paraphase amplifier, provides two outputs having the required 180-degree phase difference. The basic single-tube phase splitter is shown.

The input signal is coupled to the control grid of the stage by capacitor C_1 and resistor R_1. One of the outputs is taken from between the *cathode* and *ground* and the other is taken from between the *plate* and *ground*. The output taken from the cathode is the voltage developed across cathode resistor R_3 by the tube current.

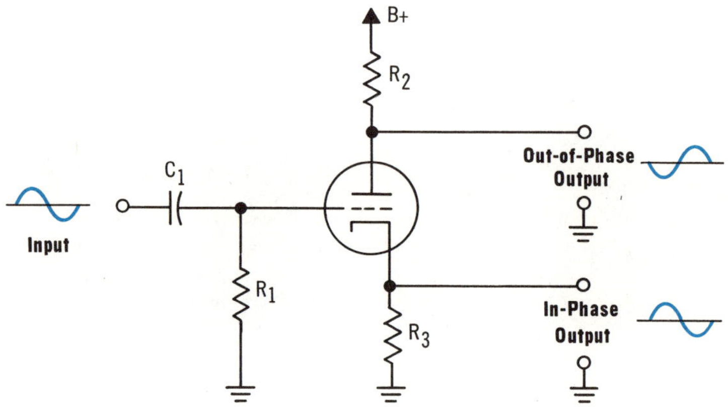

In a single-tube phase splitter, equal-amplitude but out-of-phase outputs are developed at the plate and cathode of the stage

Since cathode current increases when the input signal applied to the grid goes more positive, and decreases when the input signal goes less positive (or negative), the signal voltage developed across the cathode resistor by the cathode current is *in phase* with the input signal. The signal voltage at the plate, though, is 180 degrees *out of phase* with the input signal. The reason for this is that the signal voltage at the plate is equal to the B+ supply voltage minus the voltage drop across plate load resistor R_2 caused by the signal current. Thus, when the input signal goes more positive, plate current increases, the voltage drop across R_2 increases, and the signal voltage at the plate decreases. The converse is true when the input signal goes less positive, or negative.

Cathode resistor R_3 and plate load resistor R_2 have *equal* resistance values. Since cathode and plate current are equal in a triode, this means that both resistors produce equal voltage drops. Thus, the two output signals are of equal amplitude and 180 degrees out of phase.

the single-tube balanced-type phase splitter (cont.)

In the single-tube phase splitter, the two output signals are actually a-c variations of d-c voltage levels. The d-c component is removed by capacitive coupling

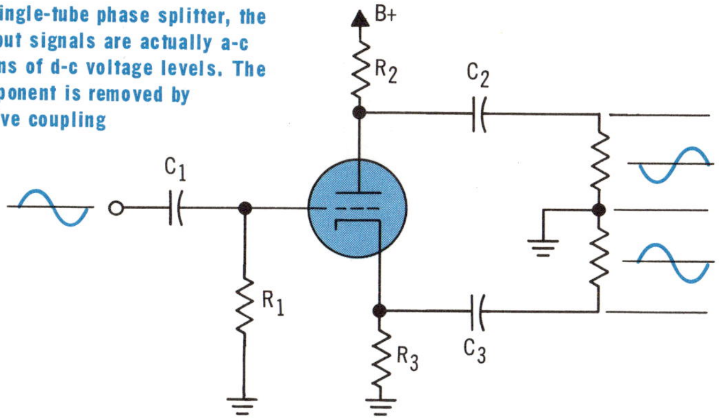

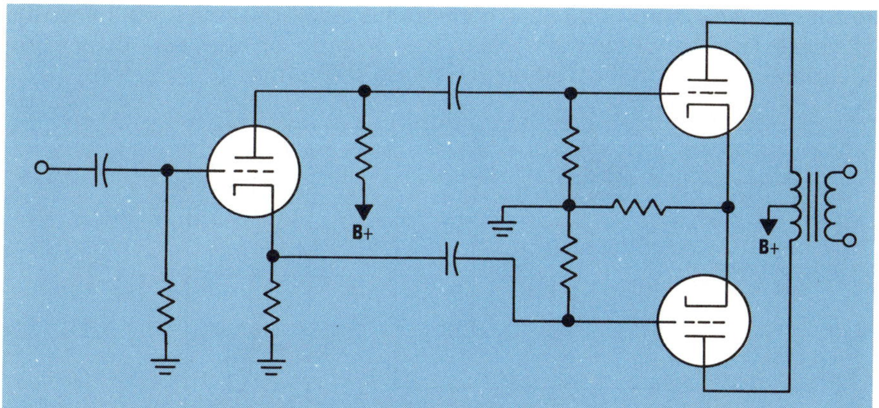

SINGLE-TUBE PHASE SPLITTER FEEDING PUSH-PULL AMPLIFIER

An important point you should understand here is that the output signals are actually the *a-c components* of the voltages appearing at the cathode and plate of the tube. These a-c components, in effect, ride on a d-c voltage level that is different at the plate than it is at the cathode. For example, with no input signal, tube current might produce a d-c voltage of 50 volts at the cathode and 250 volts at the plate. With an input signal applied, the variation in tube current might then produce voltage drops across the cathode and plate load resistors such that the voltage at the cathode varies between 40 and 60 volts, while the voltage at the plate varies between 240 and 260 volts. In both cases, the a-c component, which is the signal voltage, is 20 volts. The a-c signal component of these voltages is coupled to the next stage, which is usually a push-pull amplifier, by capacitors.

two-stage phase splitters

Single-tube phase splitters have the disadvantage of low or, actually, no gain. Each of the output signal voltages is *smaller* than the input voltage. Where amplification as well as phase splitting is desired, two-tube phase splitters are usually used.

Essentially, a two-tube phase splitter consists of *two amplifier stages.* The first stage receives the input signal, amplifies it, and inverts it in conventional amplifier style. The second stage receives as its input a *portion* of the output of the first stage. It amplifies this signal and also inverts it in the conventional way. The output of the second stage is thus an amplified version of the original input signal that has been inverted *twice.* Because of the double inversion, therefore, the output of the second stage has the same phase as the original input. Thus, the outputs of the two stages are 180 degrees out of phase.

The two basic requirements of this type of two-tube phase splitter are: (1) that the portion of the output of the first stage supplied as an input to the second stage be *equal* in *amplitude* to the original input to the first stage, and (2) that both stages amplify their inputs equally. If these two conditions are met, the outputs of the two stages are identical except for the 180-degree phase difference.

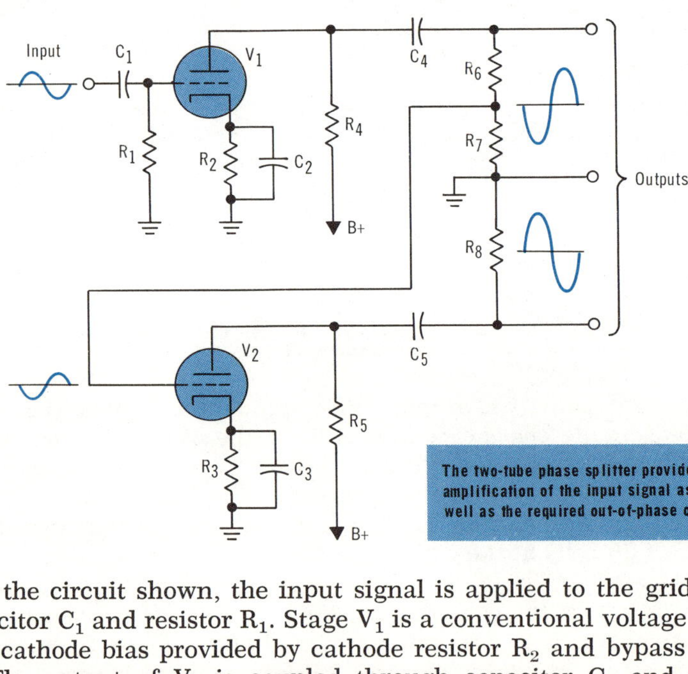

The two-tube phase splitter provides amplification of the input signal as well as the required out-of-phase outputs

In the circuit shown, the input signal is applied to the grid of V_1 by capacitor C_1 and resistor R_1. Stage V_1 is a conventional voltage amplifier, with cathode bias provided by cathode resistor R_2 and bypass capacitor C_2. The output of V_1 is coupled through capacitor C_4 and developed across resistors R_6 and R_7, which form a voltage divider. A small portion of the overall output voltage is tapped from the junction of R_6 and R_7 and applied to the grid of V_2.

two-stage phase splitters (cont.)

The input voltage to V_2 is the portion of the V_1 output signal developed across resistor R_7. The value of R_7 must be such that this small signal voltage is of the same amplitude as the input to tube V_1, which produced it. V_2 is also a conventional voltage amplifier and amplifies and inverts its input signal. The output of V_2 is coupled through capacitor C_5 and developed across resistor R_8.

The two output signals thus appear across resistors R_6 and R_7 and resistor R_8, respectively. The output across R_6 and R_7 has been inverted once, so it is 180 degrees out of phase with the original input signal. However, the output signal across R_8 has been inverted twice, so it has the same phase as the original input signal.

Another type of two-tube phase splitter was mentioned previously in the discussion of the cathode-coupled push-pull amplifier. In this circuit, the signal is eventually inverted by one stage but not by the other. The outputs of the two stages are therefore 180 degrees out of phase.

Q_1 amplifies and inverts the input signal. A portion of the output signal is taken off resistor R_7 and fed to Q_2. Q_2 amplifies and inverts this signal

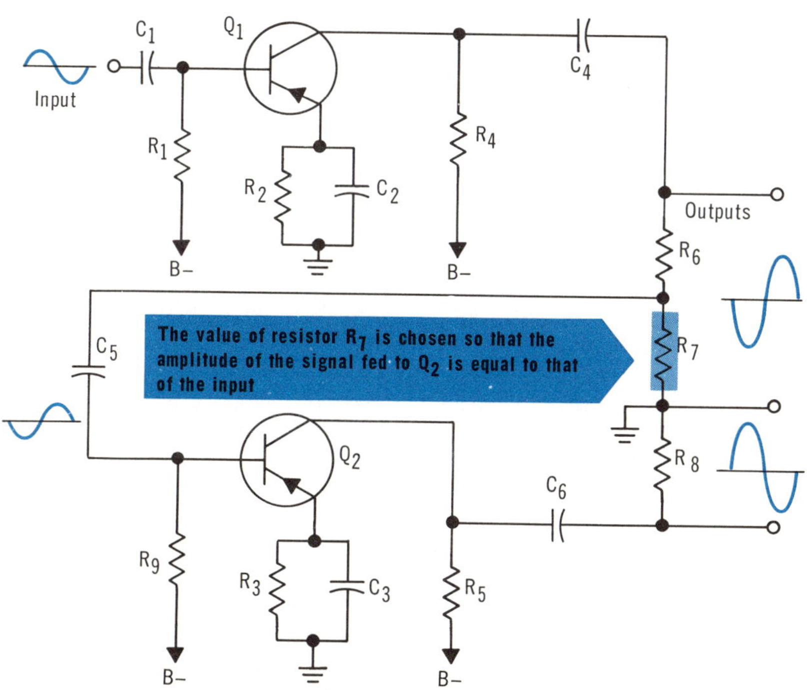

Since Q_2 amplifies and inverts the signal, its output will be equal and opposite to the output of Q_1 if Q_1 and Q_2 have the same gain

follower circuits

The term follower circuit is frequently applied to *grounded-plate amplifiers* as well as *grounded-collector amplifiers*. The term originated from the fact that in a grounded-plate amplifier the input is applied to the control grid and the output is taken from across an unbypassed cathode resistor. When an input signal is applied to such a stage, the signal current that flows in the cathode circuit produces a voltage drop across the unbypassed cathode resistor that is in phase with, or *follows*, the input signal. Thus, since the output signal at the cathode follows the variations in the input signal, such a circuit is called a *cathode follower*. Grounded-collector transistor amplifiers have characteristics similar to those of cathode followers, so they are frequently referred to as *emitter followers*.

The high input impedance and low output impedance of follower circuits make them highly suitable for impedance-matching purposes

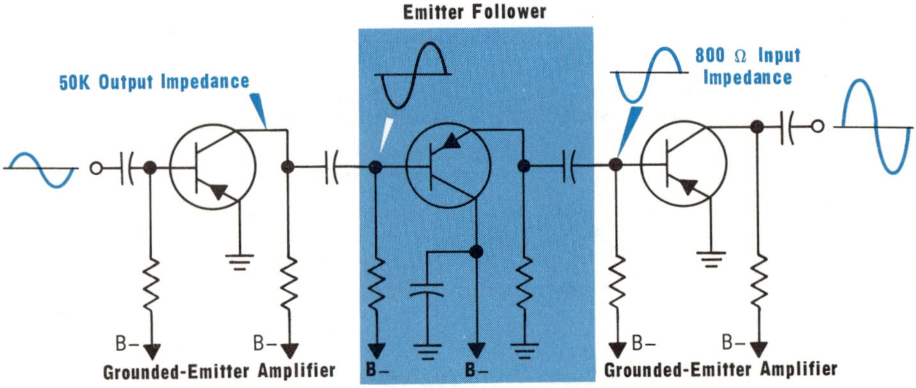

In this simplified circuit, an emitter follower is used to match the 50K output impedance of one grounded-emitter amplifier to the 800-ohm input impedance of another grounded-emitter amplifier

Other important characteristics of follower circuits besides the in-phase relationship between their inputs and outputs are their *high-input impedances* and *low-output impedances*. These impedance relationships make them very suitable for impedance-matching purposes. For example, an emitter follower can be used between two grounded-emitter amplifiers to provide a good match between the high output impedance of the first amplifier stage and the low input impedance of the second stage. Or, a cathode follower might be used to match a high-output-impedance amplifier to a low-impedance transmission line.

To obtain the good impedance-matching properties of follower circuits, voltage gain must be sacrificed. All follower circuits have a voltage gain of *less* than unity. This means that the output voltage is always slightly less than the input voltage. However, relatively large values of current or power gain can be obtained.

operation

A **Basic Cathode Follower**

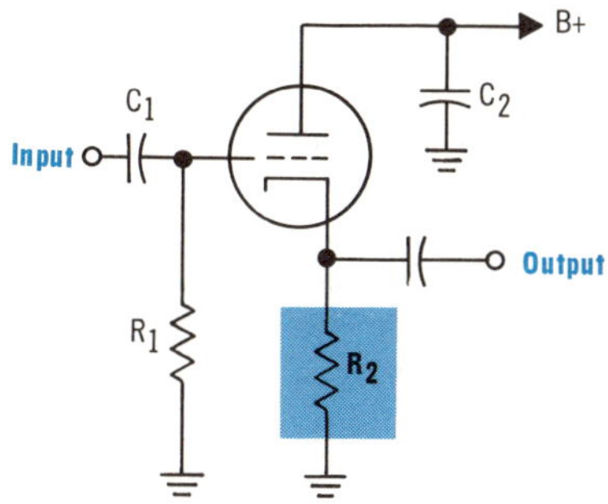

In a basic cathode-follower circuit, a single resistor is used in the cathode circuit. This single resistor develops both the output voltage and the bias for the stage

B **Modified Cathode Follower**

Limiting that occurs on the negative peaks of the input signal can be avoided by modifying the circuit so that the grid bias is reduced without affecting the overall cathode resistance necessary for a suitably high output voltage

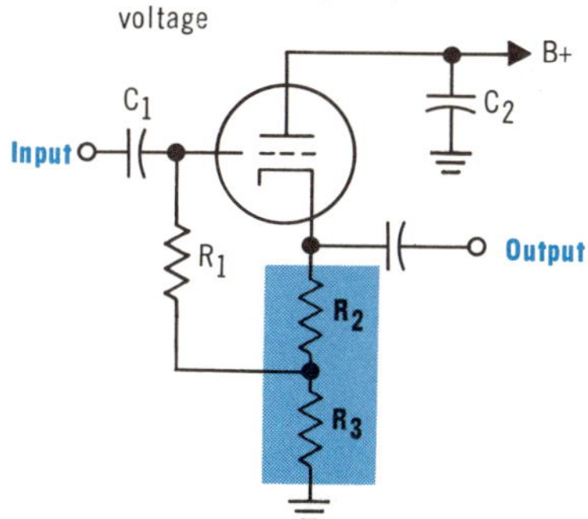

The basic cathode-follower circuit is shown in A. It essentially is a *class A degenerative amplifier*, with the plate grounded for the a-c signal by capacitor C_2 and the output taken from across unbypassed cathode resistor R_2. Since the stage is biased class A, the no-signal current flowing through R_2 establishes the normal bias for the circuit.

When an input signal is applied to the control grid, it causes the plate current to vary and produces a corresponding a-c voltage drop across cathode resistor R_2. These voltage variations make up the output signal and also vary the bias of the stage in the same direction as the input signal to cause degeneration. The degeneration, though, is responsible for the high-input impedance of the circuit, since within certain limits it prevents a large input signal from causing large changes in plate current.

Very large input signals will cause amplitude distortion of the output signal. This occurs if the input signal goes so far negative that it cuts off the tube, or so far positive that it causes the grid to draw current. If the amplitude distortion, or *limiting*, as it is called, occurs only on the negative peaks of the input signal, the modified cathode-follower circuit shown in B can be used.

In this circuit, the output signal is taken from across both of the resistors that form a voltage divider in the cathode circuit. This provides maximum output voltage. However, the grid resistor is connected to the junction of the two resistors. This reduces the grid bias by an amount equal to the voltage drop across the lower resistor, R_3. Notice that although the cathode follower is referred to as a grounded-plate amplifier, the plate does not actually have to go to d-c ground. The capacitor (C_2), though, keeps it at *a-c* or *signal ground.*

summary

☐ In a push-pull amplifier, the signal to be amplified is usually converted to two 180-degree out-of-phase signals that are applied simultaneously to the inputs of two tubes or transistors. The amplified output signal is developed across a transformer that is connected between the plates or collectors of the two stages. ☐ Push-pull amplifiers are normally operated class A, AB, or B. ☐ Cathode-coupled and emitter-coupled push-pull amplifiers can be operated with only a single input signal.

☐ Phase splitters or paraphase amplifiers are circuits that provide the separate inputs for push-pull amplifiers. A phase splitter delivers two outputs: one the same as the input, and the other 180 degrees out of phase with the input. ☐ In a transformer-type phase splitter, a transformer having a center-tapped secondary is used to produce the two out-of-phase outputs. ☐ In a single-tube paraphase amplifier, one output is taken from the cathode and the other from the plate. ☐ Two-stage phase splitters provide amplification as well as phase splitting. In a two-stage phase splitter, both stages must provide equal amplification, and the portion of the first stage output delivered to the second stage must be equal in amplitude to the input to the first stage. ☐ An inverter delivers as an output a single signal that is an inverted version of the input signal.

☐ Grounded-plate and grounded-collector amplifiers are frequently called follower circuits. ☐ Follower circuits have high input impedances and low output impedances, which make them suitable for impedance-matching purposes.

review questions

1. What is a *single-ended amplifier*?
2. What is the difference between an inverter and a phase splitter?
3. Which of these are follower circuits: a grounded-plate amplifier, a grounded-cathode amplifier, and a grounded-emitter amplifier?
4. What is a *paraphase amplifier*?
5. What is the advantage of two-stage phase splitters?
6. What is a *push-pull amplifier*?
7. Are push-pull amplifiers ever biased class C?
8. Draw a push-pull amplifier and explain its operation.
9. Do all push-pull amplifiers require two out-of-phase input signals?
10. What are the impedance characteristics of follower circuits?

video amplifiers

Any amplifier capable of uniformly amplifying signals that contain an extremely *wide range of frequencies* is considered to be a video amplifier. Generally, the frequency range of video amplifiers is from about 30 cps to several megacycles. Television, radar, and telemetering are some of the applications in which these amplifiers are widely used.

Essentially, most video amplifiers are somewhat similar to class A audio voltage amplifiers. The principal difference is that video amplifiers incorporate special *compensating circuits*. These circuits make possible the required wide frequency response. *Without* the compensating circuits, the gain of a video amplifier would be uniform over only a limited range. Frequencies above and below this limited range would be amplified very little, or not at all.

The compensating circuits used with video amplifiers are responsible for the extremely wide frequency response of these amplifiers

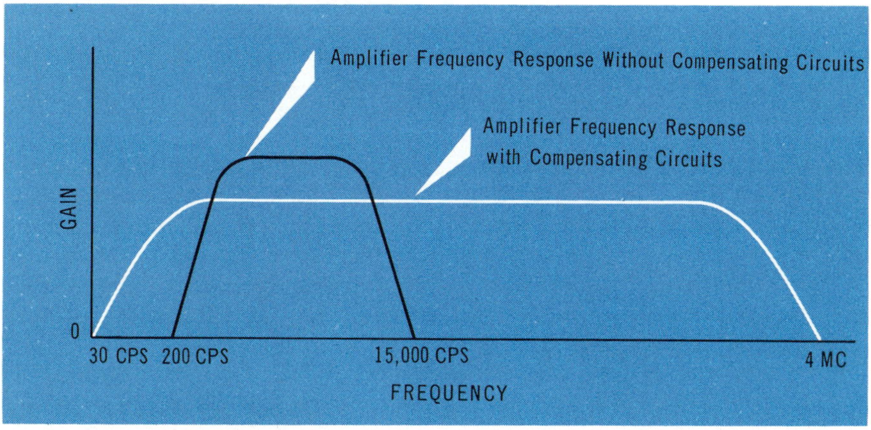

Some overall gain, however, is sacrificed to obtain the wide frequency response

Two types of compensating circuits are used with video amplifiers: those that increase the *low-frequency response* and those that increase the *high-frequency response*. These are discussed separately on the following pages.

As you will see, the wide-band requirements of video amplifiers make the selection of tubes and transistors for these circuits an important consideration. In tube-type video amplifiers, high-gain pentodes having large values of transconductance (g_m) and low values of interelectrode capacitances are generally used. In transistor circuits, good quality r-f transistors with frequency cutoffs of 100 megacycles and higher are usually used.

high-frequency compensation

A simplified video amplifier without compensating circuits is shown. In this circuit, the high-frequency response is limited by the three capacitances indicated in color. Capacitance C_o is the output capacitance of tube V_1, C_i is the input capacitance of tube V_2, and C_d is the *distributed* or stray capacitance of the circuit wiring. These three capacitances, although not physical capacitors, have the same electrical effect as if they were.

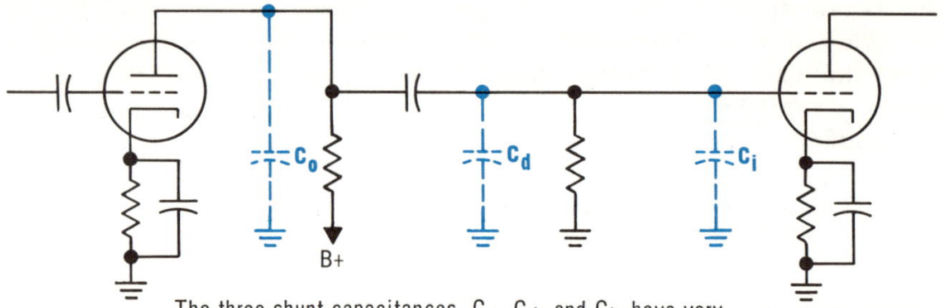

The three shunt capacitances, C_o, C_d, and C_i, have very small capacitance values. At low and medium frequencies, therefore, they have little effect on the circuit. But, since $X_C = 1/(2\pi fC)$, at high frequencies the reactance of the shunt capacitances is low

As all three of the capacitances are directly between the *signal path* and *ground*, they tend to *shunt* a portion of the signal to ground. The three capacitances have very small values, so at low and medium frequencies their reactance is very high and they have little or no shunting effect on the signal. But at higher frequencies, the reactances of the three capacitances get smaller and smaller, so more and more of the signal is shunted to ground.

The input and output capacitances (C_o and C_i) can be minimized by using tubes with low values of interelectrode capacitance. Likewise, the distributed wiring capacitance can be minimized by careful physical layout of the circuit tubes and wires. However, although they can be minimized, the three capacitances can *never* be entirely eliminated. Consequently, if the circuit is to amplify very high frequencies, some method must be used to eliminate the *effects* of these capacitances at the frequencies involved. This is the purpose of the high-frequency compensating circuits.

In transistor video amplifiers, the gain at high frequencies is limited by shunt capacitances similar to those in tube circuits. In addition, however, the high-frequency performance of a transistor depends on its alpha cutoff frequency, which is the frequency at which the current gain of a transistor connected as a grounded-base amplifier drops a definite percentage. The actual cutoff frequency depends not only on the particular transistor used, but on how it is connected (grounded-base, grounded-emitter, etc.).

circuits for high-frequency compensation

Essentially, high-frequency compensation is accomplished by using a *small inductor* to form a *parallel* resonant circuit with the shunt capacitances (*shunt compensation*), or a *series* resonant circuit with the shunt capacitances (*series compensation*).

An example of shunt compensation is shown in A. The total load impedance consists of L_1 and R_1. Capacitance C_T, which is the sum of the shunt capacitances C_o, C_d, and C_i, is effectively in parallel with L_1 and R_1. The value of peaking coil L_1 is chosen so that at the frequency at which the gain of the amplifier begins to fall off, the reactances of L_1 and C_T are equal, and they form a parallel resonant circuit. At this resonant frequency, the total plate load impedance is therefore maximum, and the output voltage is, in effect, boosted, or prevented from falling off.

In series compensation, as shown in B, a coil, L_1, is added in the signal path in series with coupling capacitor C_1. This coil forms a series resonant circuit with shunt capacitances C_d and C_i. The value of L_1 is such that at the frequency at which the amplifier gain begins to fall off, L_1 resonates with the combination of C_d and C_i. As a result, a large voltage drop is produced across C_d and C_i. Thus, the signal voltage is prevented from falling off.

Quite often, *both* series and shunt coils are used. A resistor is used in parallel with the coil to prevent a sharp resonance peak, which would amplify one frequency too much. The resistor lowers the Q of this coil to cause a very broad resonance curve that compensates for a wide band of high frequencies. Also, if the resistor were not used, a strong signal could cause the resonant circuit to produce damped oscillations, which is known as *ringing*. Usually, the coil is wound right around the resistor, and the whole unit is called a *peaking* coil.

A **Shunt Compensation**

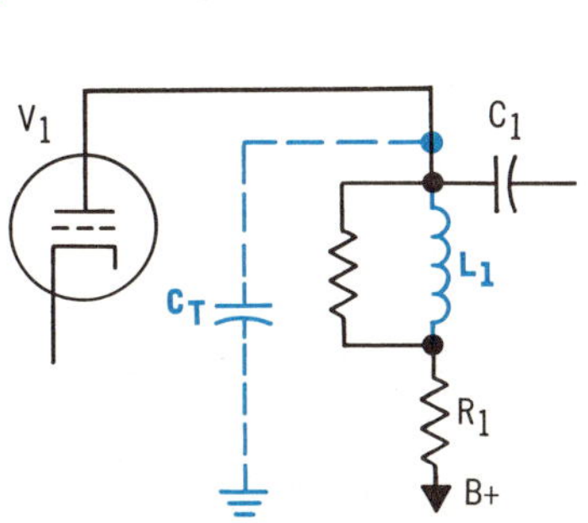

The coil connected in series with the plate load resistor forms a parallel resonant circuit with C_T, which is the total of the circuit shunt capacitances

B **Series Compensation**

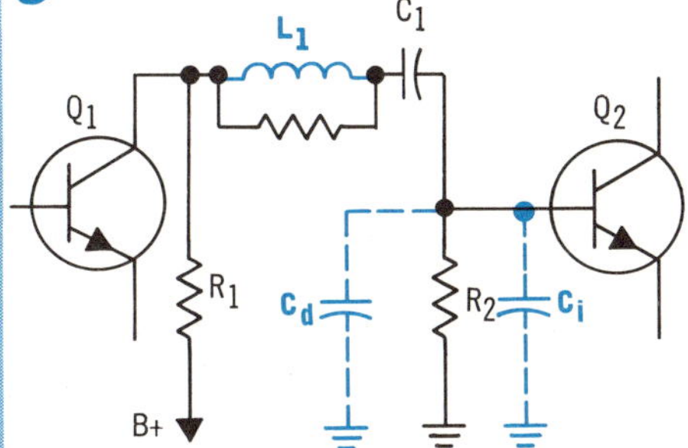

The coil connected in series with the signal path forms a series resonant circuit with the parallel shunt capacitances, C_d and C_i

other high-frequency compensation circuits

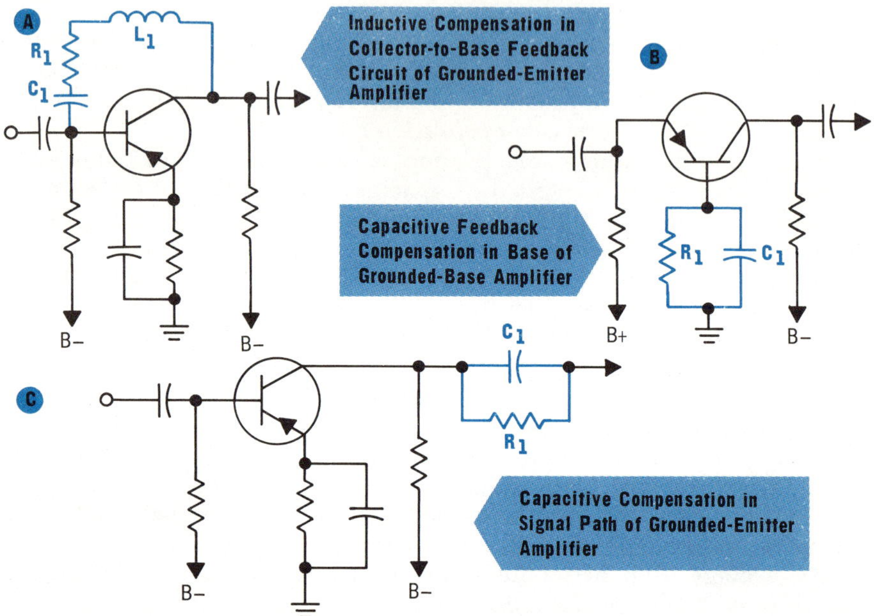

In addition to the basic series and shunt methods of compensation described on the previous page, various other methods of high-frequency compensation are often used in video amplifier circuits. One of these, called *combination* compensation, is merely the use of both series and shunt peaking coils in the same circuit. Other methods frequently used with transistor video amplifiers are illustrated in A through C.

In the circuit shown in A, the increase in the reactance of coil L_1 at high frequencies reduces the negative feedback from collector to base through resistor R_1 and coupling capacitor C_1. With less negative feedback, the gain of the stage is increased at the higher frequencies.

In the circuit shown in B, capacitor C_1 in the base circuit has a high reactance at low and medium frequencies. At high frequencies, though, its reactance decreases, thus reducing the base circuit impedance. This increases the gain of the stage at the high frequencies.

In the circuit shown in C, the parallel combination of resistor R_1 and capacitor C_1 is connected directly in the signal path. At high frequencies, where the signal level would otherwise fall off, the decrease in the reactance of C_1 allows the signal to be shunted around R_4, thus decreasing signal attenuation.

Actually, each of these methods differs from those on the previous page because those increased the high frequencies to the level of the low frequencies, whereas these methods only degenerate the lows to the level of the highs.

low-frequency compensation

The fall off in gain of a video amplifier at the low frequencies is due primarily to the *decrease* in the *reactance* of the circuit *coupling capacitor*. In addition, the low-frequency gain is also influenced by the various *bypass capacitors* used in the circuit.

In the circuit shown below, the output signal from stage V_1 is coupled through coupling capacitor C_1 and developed across grid resistor R_1. Effectively, C_1 and R_1 form a voltage divider, with the voltage appearing across R_1 being the input for stage V_2. At high frequencies the reactance of C_1 is almost zero, so for practical purposes the entire output signal from V_1 is developed across grid resistor R_1 and applied to stage V_2. At low frequencies, however, a substantial portion of the signal is dropped across the increased reactance of C_1, and the input to V_2 is decreased accordingly.

Low-frequency response is affected by the decrease in reactance of capacitors C_1, C_2, and C_3 at the lower frequencies. The values of these capacitors determine the low-frequency falloff

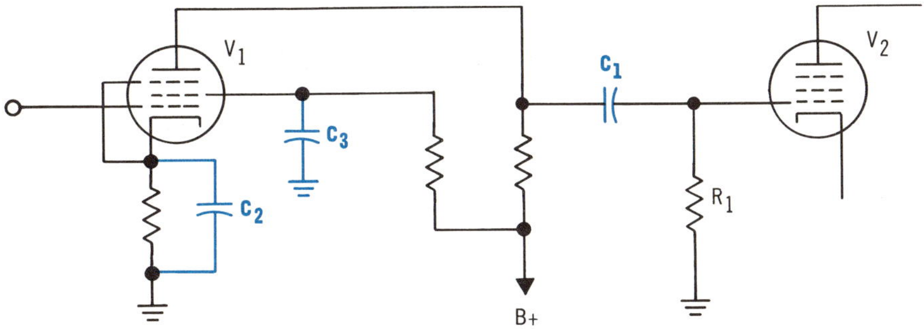

The effect of the increased reactance of C_1 at low frequencies can be avoided somewhat by increasing its capacitance value. However, if C_1 is made too large, it will start developing grid-leak bias for the stage and cause distortion. Since the signal loss across C_1 cannot be avoided, therefore, some method must be used to compensate for it.

Cathode and screen bypass capacitors C_2 and C_3, respectively, affect the low-frequency gain of the amplifier also, as a result of their increased reactance at the lower frequencies. At medium and high frequencies, the reactances of C_2 and C_3 are low enough for the capacitors to shunt the current variations around their respective resistors effectively. But at low frequencies, C_2 and C_3 no longer provide effective shunting action. As a result, the signal current variations flow through the cathode resistor and the screen current variations flow through the screen dropping resistor. Both of these conditions cause *degeneration* and, consequently, reduce the low-frequency gain of the stage.

circuits for low-frequency compensation

The signal attenuation caused by the increased reactance of a video-amplifier coupling capacitor at low frequencies can be overcome in two basic ways. One method, which is frequently used in transistor circuits, is to eliminate the coupling capacitor and use *direct coupling* between stages. When this is done, d-c bias stability is extremely important, since the separate stages are not independent in regard to dc. Any bias instabilities of the first stage could be amplified by the second stage.

In both tube and transistor circuits, the low-frequency effects of the coupling capacitor are frequently compensated for by *dividing* the load resistance and bypassing *one part* of it with a capacitor. As shown, the plate load consists of resistors R_1 and R_2, with R_2 bypassed by capacitor C_1. At high frequencies, the reactance of C_1 is low, effectively bypassing R_2, so the plate load consists of only R_1. But at low frequencies, the reactance of C_1 is large, so the plate current variations are no longer bypassed around R_2. This has the effect of inserting R_2 in the plate load circuit, since the plate current variations now flow through R_1 and R_2. With the plate load resistance thus increased, more signal voltage is developed at the plate of V_1. This increase compensates for the decrease caused by the loss of signal across the coupling capacitor.

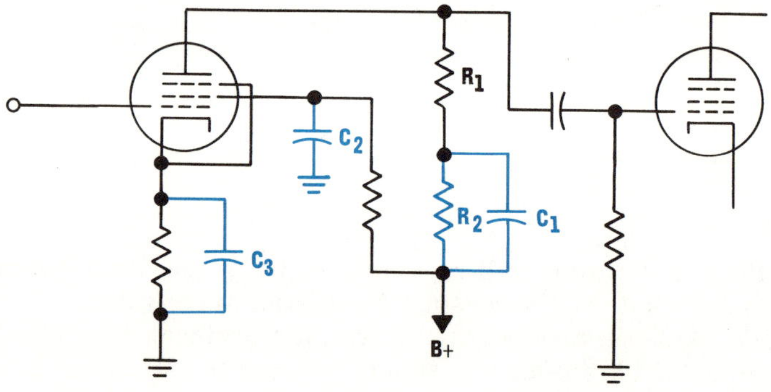

Low-frequency compensation in tube-type video amplifiers is often accomplished by using a partially-bypassed split-load resistance in the plate circuit and large value screen and cathode bypass capacitors

To compensate for the degeneration caused by the decreased reactance of bypass capacitors, video amplifiers use *large-value capacitors* for bypass purposes. These capacitors usually have values many times larger than those used in comparable audio amplifiers. In this way, even at low frequencies the capacitor reactance is low enough to provide the required bypass action and avoid degeneration.

r-f amplifiers

R-f amplifiers could be defined as circuits that amplify voltages and currents having frequencies that lie in the r-f portion of the frequency spectrum. However, this definition is not very satisfactory, since there is no definite agreement as to just what frequencies make up the r-f portion of the frequency spectrum. A better definition of r-f amplifiers is based on their *function,* which is to be able to select and amplify a *narrow band* of high frequencies.

Only a very narrow band of frequencies is amplified by an r-f amplifier at one time. The center frequency of this narrow band, though, can be varied by changing the resonant frequency of the tuned circuits that make up part of the r-f amplifier

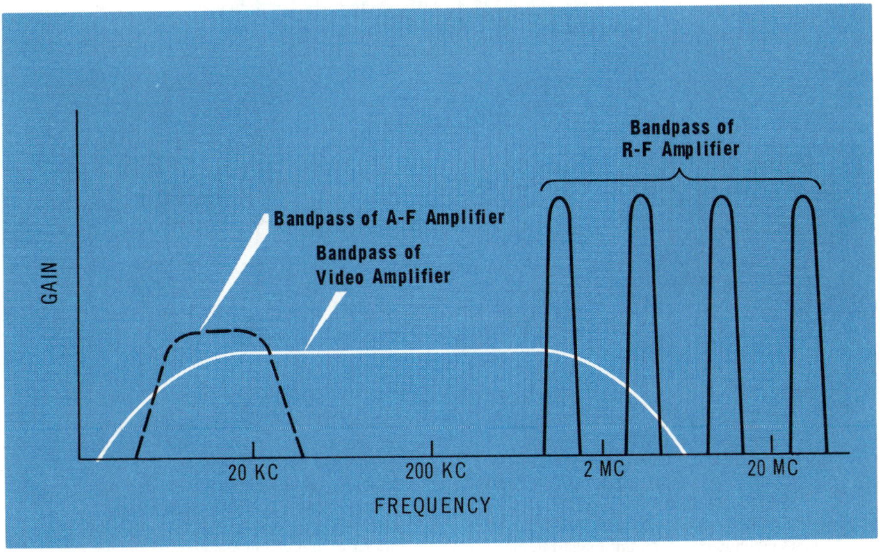

Thus, as shown above, while a-f and video amplifiers amplify only a fixed band of frequencies, the frequencies to be amplified by an r-f amplifier can be varied

Generally, r-f amplifiers are characterized by *tuned,* or resonant, circuits in their inputs and outputs. The resonant frequency of these tuned circuits can usually be changed.

There are various types of r-f amplifiers, which, although basically similar, have definite differences. Because of these differences, each type of r-f amplifier will be covered separately on the following pages. These various types include voltage amplifiers, buffer amplifiers, frequency multipliers, and power amplifiers.

r-f voltage amplifiers

R-f voltage amplifiers, as their name implies, simply amplify the signal voltages applied to their inputs. They are generally operated class A and have plate load impedance values that produce large output voltages. In these respects, they are similar to audio voltage amplifiers. However, audio voltage amplifiers are designed to amplify equally *all* signal voltages having frequencies within the desired frequency response range of the amplifier. R-f voltage amplifiers, on the other hand, are designed to select and amplify only a *narrow band* of frequencies, and to reject, or suppress, all frequencies outside of this band.

One type of basic r-f amplifier uses transformer coupling and tuned circuits in both its input and its output

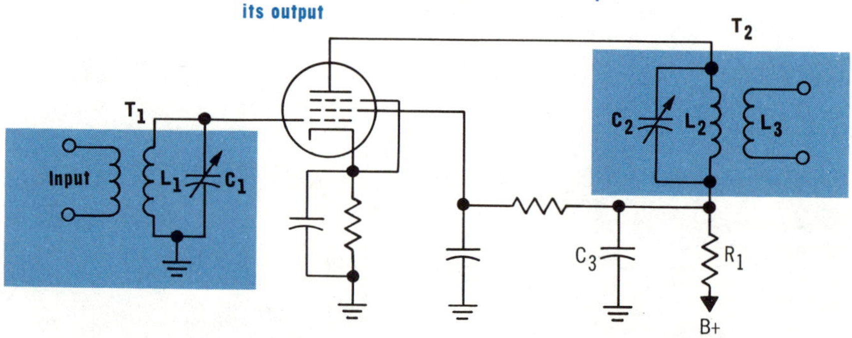

The input signal is transformer coupled to the tuned circuit consisting of transformer secondary L_1 and capacitor C_1. Maximum grid input signal is developed at the tuned frequency; little or no signal at other frequencies. The output signal is developed across the tuned circuit consisting of L_2 and C_2. Again, the maximum signal output is developed at the tuned frequency

Capacitors C_1 and C_2 are variable and are adjusted so that both tuned circuits resonate at the same frequency. Resistor R_1 and capacitor C_3 in the plate circuit of the stage form a decoupling network. R_1 offers a high impedance to the r-f component of the plate current and C_3 bypasses the r-f to ground. This keeps the r-f out of the B+ supply

This frequency *selectivity* is accomplished by tuned circuits that form part of the input or output (or both) of the amplifier. If there is a tuned circuit in the input, it passes those frequencies in the narrow band to which it is tuned to the amplifier input and rejects all other frequencies. If there is a tuned circuit in the output, it generally presents a large output load impedance at its tuned frequency and a low impedance at all other frequencies. Since the amplitude of the output signal depends on the value of output impedance, large outputs are delivered only at the tuned frequency.

r-f voltage amplifiers (cont.)

You can see that if tuned circuits are used in both the input and output of the amplifier, they must both be tuned to the *same frequencies*. When a new band of frequencies is to be amplified, both circuits must be retuned to the new frequencies.

It should be mentioned here that it is common practice to speak of an r-f amplifier or a tuned circuit as being tuned to *one* frequency, such as 1200 kilocycles or 27 megacycles. This single frequency is the *resonant frequency* of the tuned circuit, or circuits, involved. However, no practical tuned circuit can pass a single frequency and suppress all others. Some frequencies on both sides of the resonant frequency are always passed with essentially the same amplitude as the resonant frequency. These frequencies of significant amplitude make up the pass band, or *bandpass*, of the tuned circuit.

Tuned circuits used with transistor r-f amplifiers must not only develop maximum signal at the tuned frequency, they must also match the transistor input and output impedances. The method shown here, in which the tuned circuits are tapped, is often used

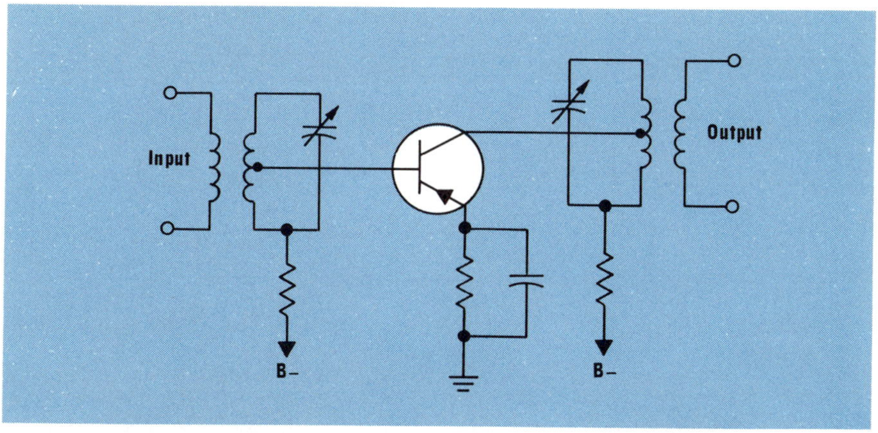

As you probably realize by now, the study of r-f amplifiers is in large part a study of tuned circuits. For this reason, a summary of tuned circuits and their properties is given on the following pages.

Transistor as well as tube circuits are covered on the following pages. Strictly speaking, the transistor circuits are not voltage amplifiers, since, as you know, all transistors are current-sensitive devices. Instead, the transistor amplifiers covered under r-f voltage amplifiers really come under the category of *small-signal* amplifiers, as opposed to power-amplifier transistor circuits. However, to avoid the necessity of continually switching terminology, both small-signal, transistor r-f amplifiers and tube-type, r-f voltage amplifiers will be covered under the heading of r-f voltage amplifiers.

the tuned circuit

A tuned circuit, as you know, is merely another name for a resonant circuit. From your studies of basic electricity, you recall that there are two types of resonant circuits: *series* resonant circuits and *parallel* resonant circuits. Both types consist of an inductance (L) and a capacitance (C) with the two elements connected in series in a series resonant circuit and in parallel in a parallel resonant circuit.

An important point for you to understand here is that the series or parallel arrangement of the inductance and capacitance in resonant circuits is with respect to the *applied voltage*. If the input voltage is applied across both the L and C components in parallel, it is a parallel resonant circuit. If the voltage is applied in series with L and C, it is a series resonant circuit. Thus, as shown, if the L component of a resonant circuit is the secondary of a transformer, the circuit is a series resonant circuit. This is because the transformer secondary voltage is induced *within* the secondary winding, which is the input to the resonant circuit. It acts as if it were an a-c generator in *series* with the secondary winding.

In a PARALLEL RESONANT CIRCUIT . . .

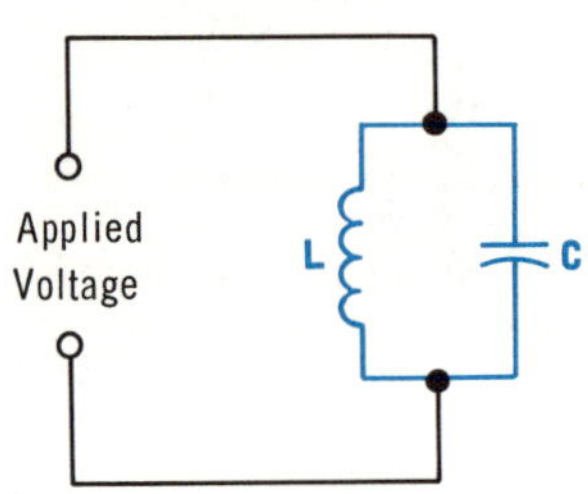

. . the voltage is applied across L and C in parallel

In a SERIES RESONANT CIRCUIT . . .

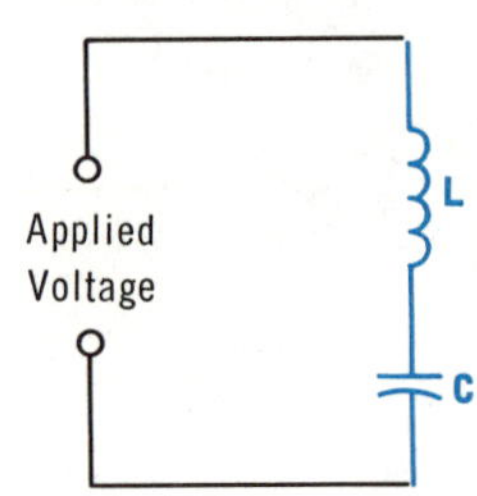

. . . the voltage is applied in series with L and C

This is a SERIES RESONANT CIRCUIT . . .

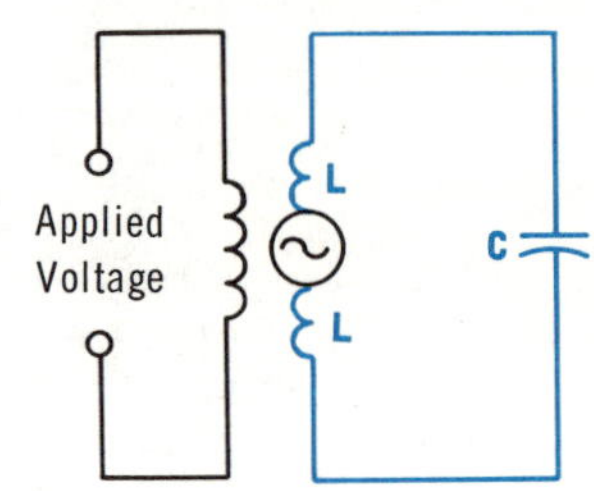

. . . because the applied voltage is effectively in series with L and C

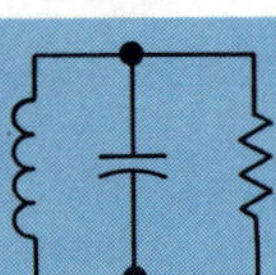

By shunting the tank circuit with a resistor, you can lower the Q and broaden the bandpass

In both series and parallel tuned circuits, resonance occurs when the *reactances* of the inductive and capacitive elements are *equal*. For any combination of inductance and capacitance, this occurs at only a *single* frequency, f_R, which can be calculated from the equation:

$$f_R = \frac{1}{2\pi\sqrt{LC}}$$

the tuned circuit (cont.)

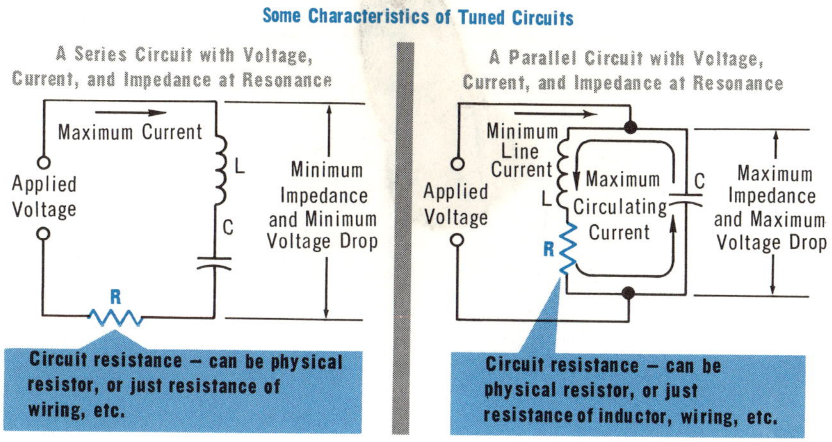

At the resonant frequency, a *series* resonant circuit has *minimum impedance* and *maximum current;* a *parallel* resonant circuit at the resonant frequency exhibits *maximum impedance* and *minimum line current.* This minimum line current is the current entering and leaving the parallel resonant circuit.

Within the tank circuit, as it is called, large values of *circulating current* flow as a result of the alternate storage and release of electrical energy by the magnetic field of the inductance and electrostatic field of the capacitance. At frequencies above and below resonance, the impedance of a series resonant circuit increases, while its current decreases. Similarly, above resonance, the impedance of a parallel resonant circuit decreases and line current increases. The degree with which these conditions change at frequencies above and below resonance is a measure of the circuit's ability to *discriminate* between frequencies.

This ability to discriminate between frequencies is called the Q of the circuit, and can be calculated from the equation:

$$Q = X_L/R \quad \text{or} \quad X_C/R$$

where X_L and X_C are the respective, and equal, reactances at the resonant frequency, and R is the resistance of the circuit. You can see from this that the larger the value of the circuit resistance, the lower the Q of the resonant circuit, and, therefore, the less able it is to discriminate between various frequencies.

Some of the more important characteristics of tuned circuits are illustrated on the following page. In some circuits, where we want to widen the bandpass, or make the tank less selective, say to make sure that all signal side bands are passed, we lower the Q of the tank by shunting it with a resistor of the proper value.

characteristics of the tuned circuit

Effect of Q on Response ($Q = X_L/R$)

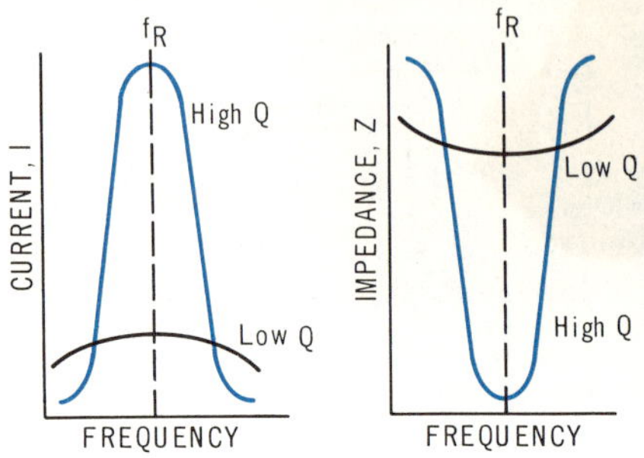

Effect of Q on Response ($Q = X_L/R$)

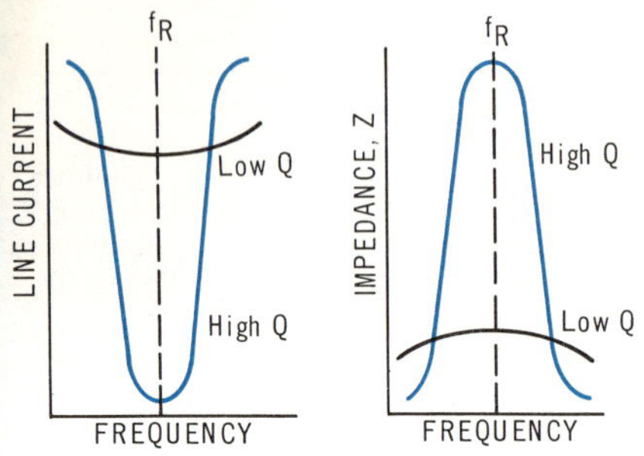

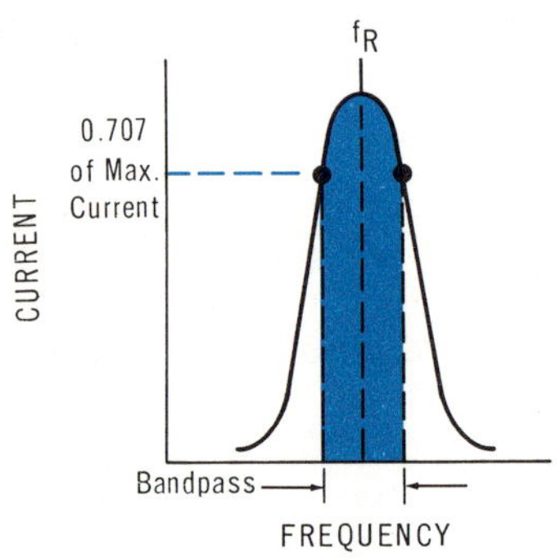

In a series tuned circuit, the bandpass is the frequency range between the two frequencies at which the circuit current is 0.707 times its value at resonance. The bandpass can be calculated from the equation: Bandpass (CPS) $= \frac{f_R}{Q}$

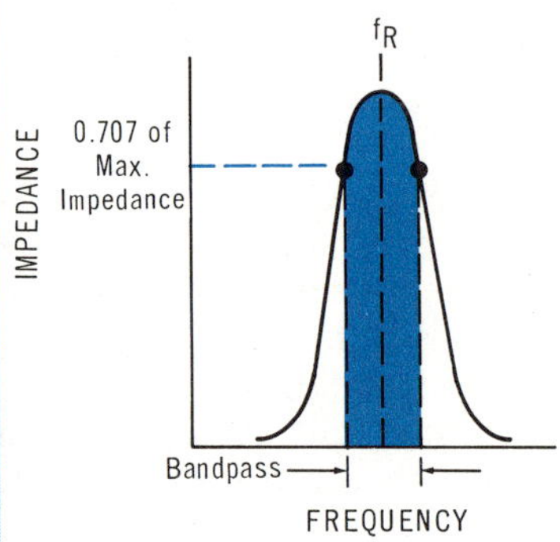

In a parallel tuned circuit, the bandpass is the frequency range between the two frequencies at which the circuit impedance is 0.707 times its value at resonance. The bandpass can be calculated from the equation: Bandpass (CPS) $= \frac{f_R}{Q}$

The Resonant Frequency of Tuned Circuit Can Be Fixed or Variable

Fixed Tuned Circuits

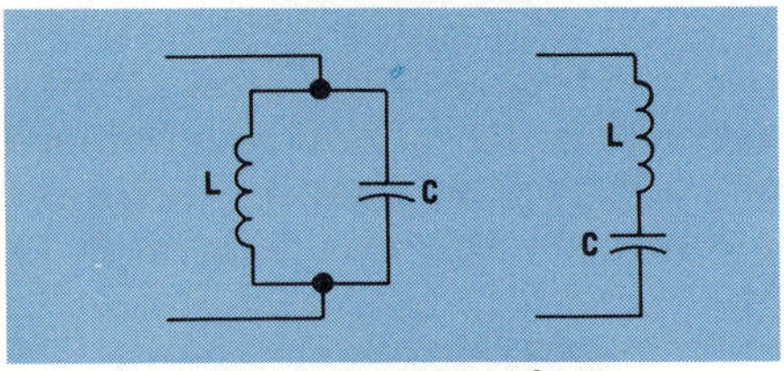

If the values of L and C are fixed, the circuit has one resonant frequency, which cannot be changed

Variable Tuned Circuits

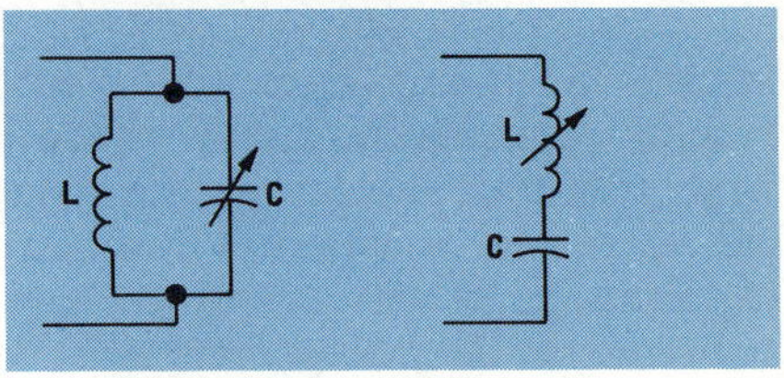

If either L or C, or both, are variable, the resonant frequency of the circuit can be changed

r-f amplifier tuned circuits

A simplified, tube-type r-f amplifier stage is shown in A. The input signal to this circuit comes from an *untuned* antenna circuit consisting of a receiving antenna and antenna coil L_1. The antenna coil forms the primary of a transformer (T_1), whose secondary is the inductive element of a *series* resonant circuit made up of L_2 and C_1.

A

In this r-f amplifier, a series resonant circuit is used in the input and a parallel resonant circuit is used in the output

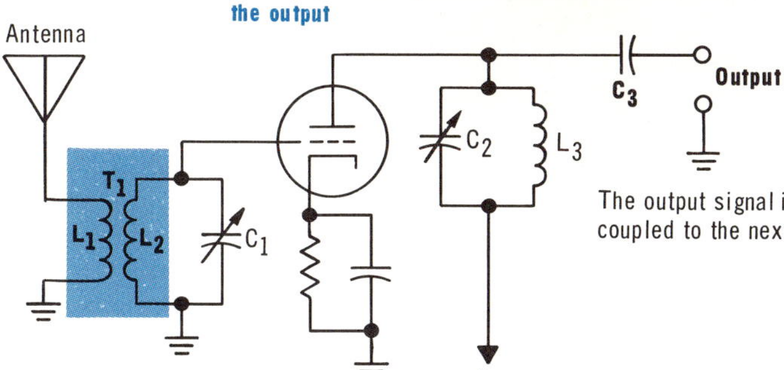

The output signal is capacitively coupled to the next stage

The signal is transformer-coupled to the input circuit

The output signal can also be transformer-coupled to the next circuit using . . .

B

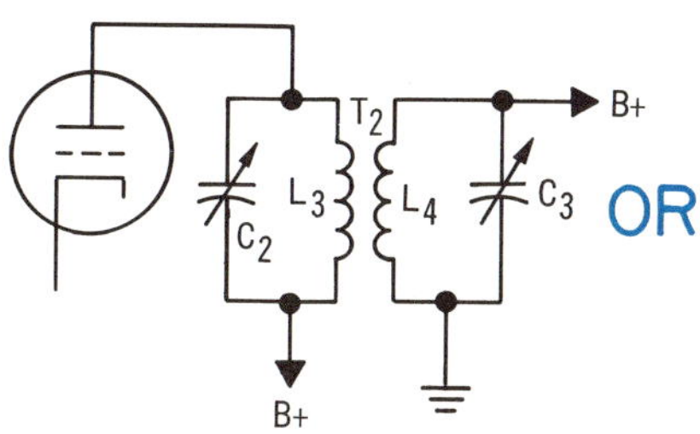

OR

C

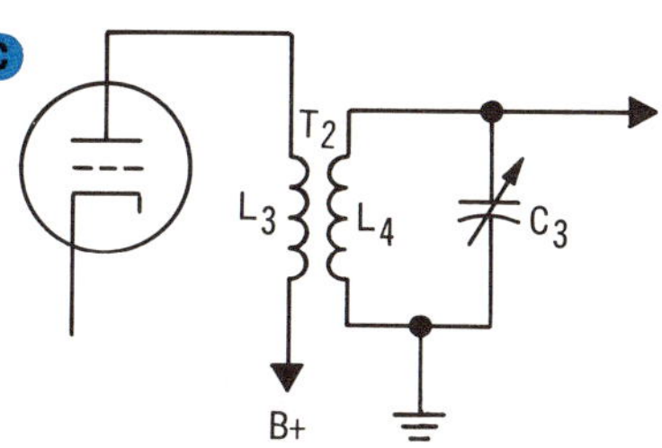

a parallel resonant circuit to transformer-couple the signal to series resonant circuit. This is frequently done in i-f amplifiers

only a coil as the plate load impedance and a series resonant circuit as the input to the next stage

Although signal currents of many frequencies flow through the antenna coil, only at the resonant frequency of L_2 and C_1 is an appreciable signal coupled to the tuned circuit. This is because at all but the resonant frequency the tuned circuit offers a high impedance. At the resonant frequency, though, a *large signal current* flows in the tuned circuit, creating a substantial voltage drop across capacitor C_1. This voltage drop follows the signal variations and serves as the grid input to the stage. Capacitor C_1 is variable and permits the resonant frequency of the tuned circuit and, therefore, the frequency of the selected signal to be changed.

r-f amplifier tuned circuits (cont.)

The output load impedance of the stage is provided by the *parallel* resonant circuit consisting of coil L_3 and capacitor C_2. At the resonant frequency, this tuned circuit offers a high impedance to the a-c component of the plate current, so maximum output voltage is developed. To signal currents that are not of the resonant frequency, the parallel combination of C_2 and L_3 offers a low impedance, so little or no output voltage is developed at these frequencies. The output signal is coupled to the next stage by coupling capacitor C_3.

Instead of capacitive coupling, the output could also be transformer-coupled to the next stage. Two ways in which this is frequently done are shown in B and C on the preceding page. In B, the output signal is transformer-coupled by the parallel resonant circuit to a series resonant circuit in the input of the next stage. The coils of the two resonant circuits make up the primary and secondary of the coupling transformer. This is called *double-tuned coupling* and is used very often for i-f amplifiers, as you will learn later.

Quite often, at the higher frequencies, a capacitor is not needed to resonate with the coil; there is enough capacitance between the windings of the coil to produce a tuned circuit

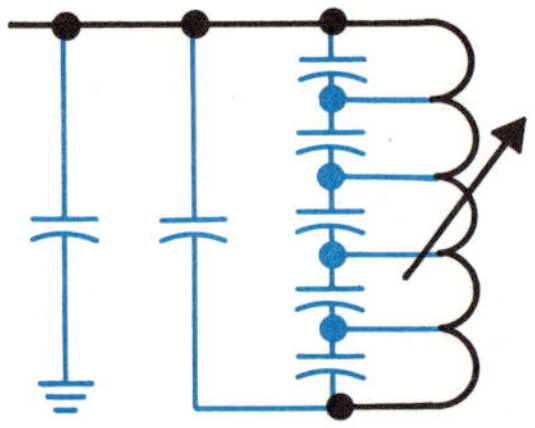

The coil is tuned with a slug to resonate with the interwinding capacitance

In C, the parallel resonant circuit is eliminated and only a coil is used as the plate load. This coil serves as the primary of the transformer for coupling the signal to a tuned circuit in the input of the next stage. This is called *single-tuned coupling*. The particular type of coupling used with r-f amplifiers depends on the desired characteristics, such as gain, bandwidth, and selectivity. You will learn more about this later.

As shown above there is usually enough capacitance between the windings of a coil at the higher frequencies so that a capacitor is not needed to form a tuned circuit. The coil then resonates with its own interwinding capacitance. The coil is tuned with a powdered iron or brass slug that is moved in and out of it.

tuned circuits for transistor r-f amplifiers

The tuned circuits used in the inputs and outputs of transistor r-f amplifiers perform the same function that they do in tube circuits. In other words, they make possible the selection of the desired signal frequencies and the rejection of all other frequencies. However, in transistor circuits the necessity of *matching* the input and output impedances of the transistors to the impedances of the tuned circuits creates problems in the design of the tuned circuits. This is especially true of the tuned circuits used as inputs to grounded-emitter and grounded-base amplifiers, which have *very low input impedances*. If such tuned circuits were of the same design as those of comparable tube circuits, they would require impractical and inefficient component values.

To overcome this problem, the tuned circuits used with transistor r-f amplifiers are designed for efficient tuning, and are then *tapped,* so that, in effect, the transistor "sees" the required lower value of impedance. A typical circuit of this kind is shown in A. By tapping the coil at the proper point, the required low impedance is matched to the transistor input. Other ways in which a tuned circuit can be tapped to provide lower impedance values are shown in B through D.

A

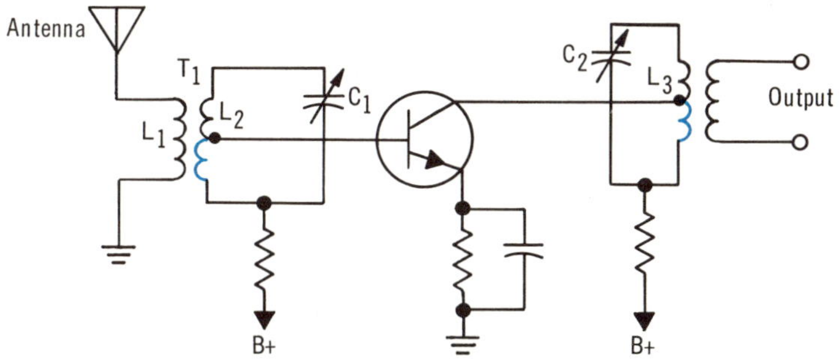

By tapping the tuned circuits, the transistor uses only the impedances of the lower portions of the tuned circuit coils. The impedances of these portions of the coils match the transistor input and output impedance

These are some other methods used to tap the tuned circuits to accomplish the same impedance matching results

C

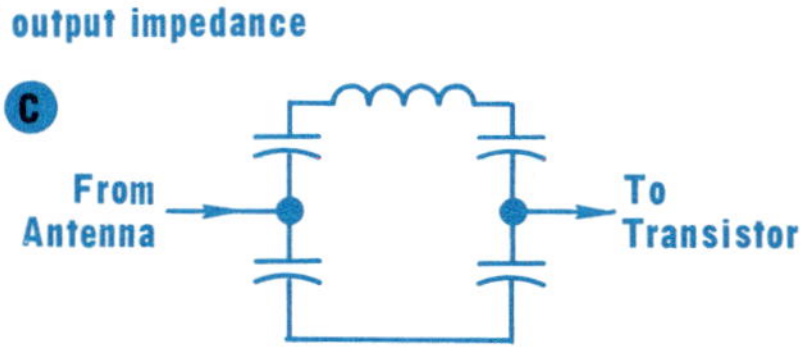

B

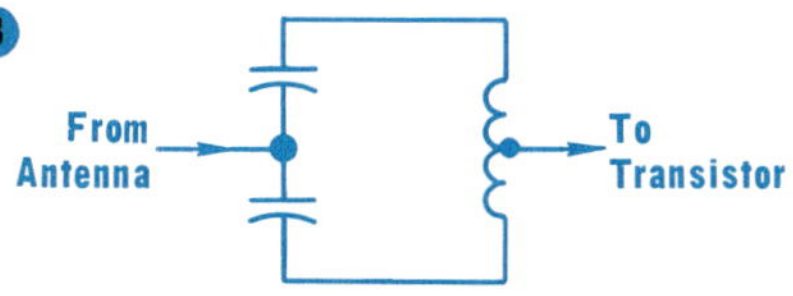

D

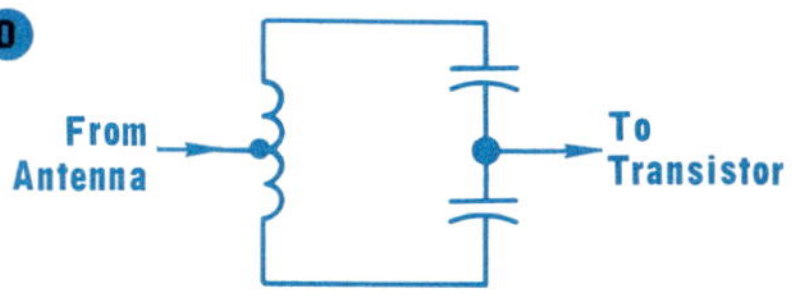

summary

☐ High-frequency compensation is required in video amplifiers because of the shunting effect caused by the input and output capacitances of the circuits and the distributed or stray capacitance of the circuit wiring. ☐ High-frequency compensation circuits prevent the output of the amplifier from falling off at high frequencies. ☐ Low-frequency compensation is required in video amplifiers because of the increase in circuit reactances, especially that of the coupling capacitor. The low-frequency compensation circuits prevent the output of the amplifier from falling off at low frequencies. ☐ In the transistor video amplifiers, low-frequency compensation is often accomplished by eliminating the coupling capacitor and using direct coupling between stages. ☐ Large-value bypass capacitors are used in video amplifiers for low-frequency compensation.

☐ R-f amplifiers are able to select and amplify a narrow band of high frequencies. ☐ Frequency selectivity of r-f amplifiers is accomplished by tuned, or resonant, circuits that form part of the input or output circuit of the amplifier.

☐ In both series and parallel tuned circuits, resonance occurs when the reactances of the inductive and capacitive elements are equal. ☐ At resonance, a series resonant circuit has minimum impedance and maximum current. ☐ A parallel resonant circuit at the resonant frequency exhibits maximum impedance and minimum line current. ☐ The Q of a tuned circuit is a measure of its ability to discriminate between frequencies. The formula for Q is $Q = X_L/R$ or X_C/R. ☐ At high frequencies, a coil can resonate with its own interwinding capacitance and thus act as a tuned circuit.

review questions

1. Why is low-frequency compensation required in video amplifiers?
2. Why is high-frequency compensation required in video amplifiers?
3. What are some of the methods used for low-frequency compensation?
4. What is a *peaking coil*?
5. How does the bandwidth of an r-f amplifier compare with that of a video amplifier?
6. What is a *tank circuit*?
7. At resonance, what is the impedance of a tank circuit?
8. What is *double-tuned coupling*?
9. Does a tank circuit always require a capacitor?
10. What is the Q of a tuned circuit a measure of?

bandwidth and selectivity

The bandwidth of an r-f amplifier is the range between the two frequencies at which the amplified output voltage falls to a value 0.707 times its value at the frequency to which the amplifier is tuned. Thus, if the output voltage of an amplifier tuned to 500 kilocycles is 100 volts at 500 kilocycles, and the output falls to 70.7 volts at 495 kilocycles as well as 70.7 volts at 505 kilocycles, the bandwidth of the amplifier is 505 − 495 kilocycles, or 10 kilocycles.

You will recognize that the definition of amplifier bandwidth is essentially the same as that of the bandwidth of a simple tuned circuit, which was discussed previously. This is natural, of course, since tuned circuits provide an amplifier with its frequency-discriminating ability, and, hence, its bandwidth.

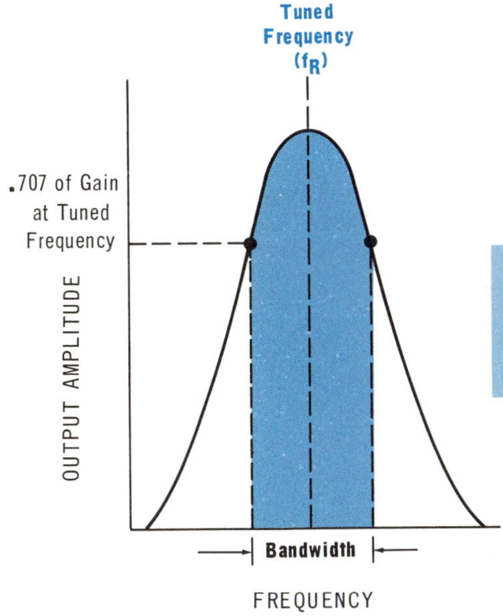

The bandwidth of a tuned r-f amplifier lies between the two points on the output amplitude vs. frequency curve at which the output amplitude is .707 times its maximum value

The bandwidth of an r-f amplifier determines how well it can select and amplify a desired signal and reject all other signals, even those very close in frequency to the desired one. An amplifier with a *narrow bandwidth* has *good selectivity.* This means that it can select the desired signal from other signals close to it in frequency. These adjacent channel signals are greatly attenuated (not amplified) so that they do not exist in the amplifier output. An amplifier with a *wide bandwidth* has *poor selectivity.* Although it amplifies the desired signal, it also amplifies with appreciable amplitude some of the adjacent channel signals. The adjacent channel signals therefore appear in the amplifier output along with the desired signal, resulting in distortion of the desired signal.

bandwidth and selectivity (cont.)

It may seem that it is desirable for an r-f amplifier to have as narrow a bandwidth as possible, since selectivity is directly related to bandwidth. This is not necessarily always the case, however. If the amplifier is to receive *modulated signals,* its bandwidth must be wide enough so that it will equally amplify all of the frequencies contained in the modulation. In the case of a conventional broadcast AM signal, this means that the r-f amplifier must have a bandwidth of at least 10 kilocycles, since the side bands of such signals cover a 10 kilocycle range. A bandwidth of less than 10 kilocycles would result in severe amplitude distortion of the amplified signal.

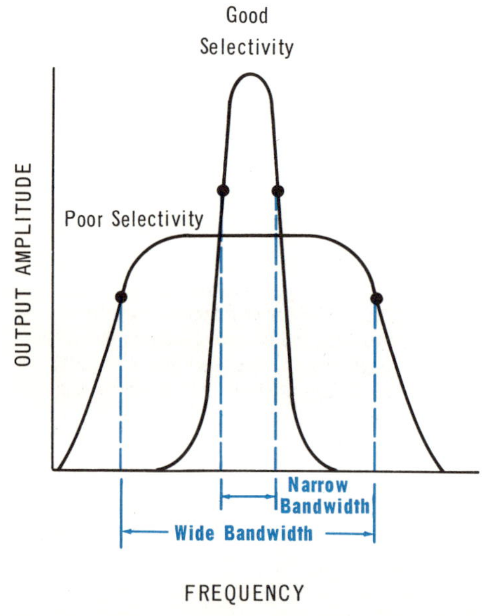

An r-f amplifier with a narrow bandwidth has good selectivity. One with a wide bandwidth has poor selectivity

For amplifiers that have single-tuned circuits, the bandwidth is equal to:

$$\text{Bandwidth} = f_R/Q$$

where f_R is the frequency to which the amplifier is tuned and Q is the Q of the tuned circuits used. You will recognize this as the equation for the bandwidth of a simple resonant circuit. If double-tuned circuits are used, this equation no longer holds, since the degree of magnetic coupling between the two tuned circuits then plays an important role in determining bandwidth. As was pointed out previously, double-tuned circuits are used more with i-f amplifiers than with r-f amplifiers. The effect of the degree of coupling on the amplifier bandwidths will therefore be covered later in this volume in the discussion of i-f amplifiers.

triodes vs. pentodes

Any *noise* introduced by the r-f stage is amplified together with the signal and tends to reduce the *signal-to-noise ratio* of the amplified output.

In tube circuits, the type of tube used has a significant effect on the noise generated by a stage. Triodes, because they have fewer grids, are inherently less noisy than are pentodes. From a noise standpoint, therefore, they are frequently desirable for r-f amplifiers. However, pentodes provide greater gain than do triodes, so from a gain standpoint, pentodes are often preferred.

In addition to affecting the noise introduced by the amplifier, the choice of a triode or a pentode also affects the circuit configuration of the amplifier. When *pentodes* are used, the circuit is usually a *grounded-cathode amplifier,* since this type of circuit provides the largest voltage gain. Triode r-f amplifiers, on the other hand, are not generally used in the grounded-cathode configuration, particularly when tuned circuits are used in both the input and output. The reason for this is that such circuits tend to oscillate like tuned-plate, tuned-grid oscillators as a result of positive feedback through the plate-to-grid interelectrode capacitance of the tube. If triodes are used in this way, *neutralization* must be employed to eliminate the feedback. This is covered on the following page. Oscillation is covered in Volume 6.

More commonly, triodes are used as *grounded-grid r-f amplifiers.* In this way, positive feedback and oscillation are not a problem, since the input signal is introduced into the cathode circuit. However, grounded-grid amplifiers, because of their low input impedance, do not have as large a voltage gain as grounded-cathode amplifiers. Because of this, it is sometimes necessary to use additional stages of r-f amplification.

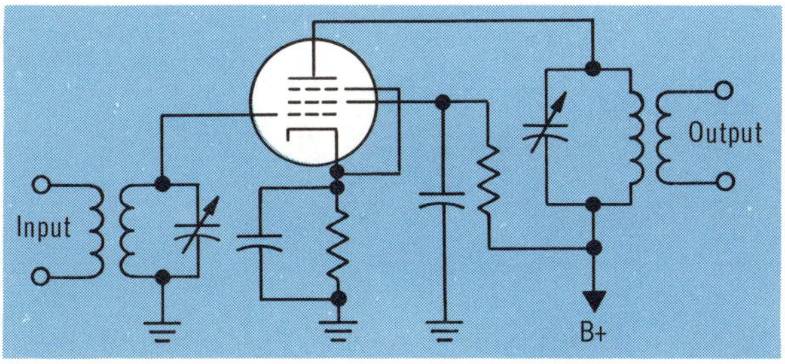

Pentodes are generally used as grounded-cathode r-f amplifiers. If triodes are used in this way, neutralization is necessary

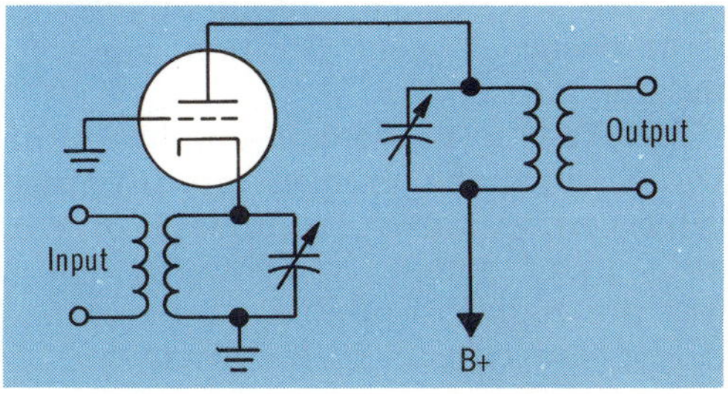

Triodes are commonly used as grounded-grid amplifiers. The relatively low gain of both the triode and the grounded-grid configuration often make additional stages of r-f amplification necessary

neutralization

You learned on the previous page that when a triode is used in a grounded-cathode amplifier having tuned circuits in both the input and output, positive feedback can occur, and the circuit may tend to function as a tuned-plate, tuned-grid oscillator. This effect becomes especially significant at signal frequencies above about 500 kilocycles, since at these high frequencies the reactance of the tube's interelectrode capacitance feedback path is relatively low.

In transistor circuits, a similar situation exists because of the internal feedback caused by the transistor's *collector-to-base capacitance*. Neutralization is essentially a process of *externally* feeding back from output to input a signal that is almost *equal in amplitude* to but *opposite in phase* with the unwanted internal feedback. Strictly speaking, in transistor circuits neutralization does not completely cancel the unwanted feedback signal. Instead, it only cancels enough of the feedback to ensure stable operation.

The triode can be neutralized in the following manner

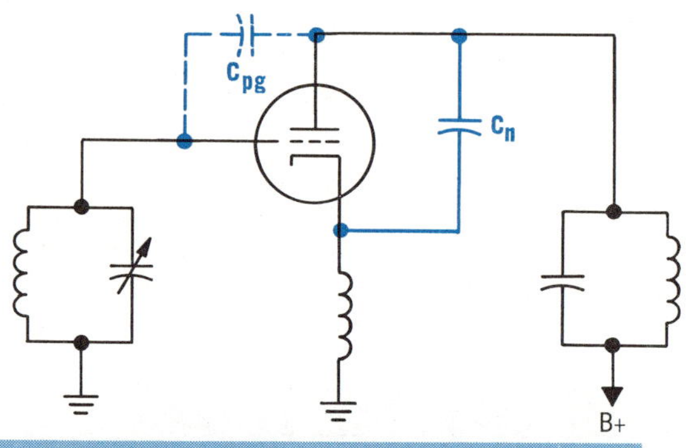

Interelectrode capacitance C_{pg} feeds back a signal that tends to make the amplifier work like an oscillator. But, when capacitor C_n is added to feed back a degenerative signal to the cathode, the feedback from plate to grid is cancelled

When the external feedback completely cancels the internal feedback, the process is called *unilateralization*. In this volume, we will only cover transistor neutralization, since it is used more commonly than unilateralization.

In tube-type r-f voltage amplifiers, neutralization is not always employed, since the problem of unwanted feedback can be avoided by using certain tubes (pentodes) or circuit configurations (grounded-grid). In r-f power amplifiers, though, grounded-cathode triode amplifiers are highly desirable, so neutralization is used extensively. This will be covered later in the discussion of r-f power amplifiers.

neutralization (cont.)

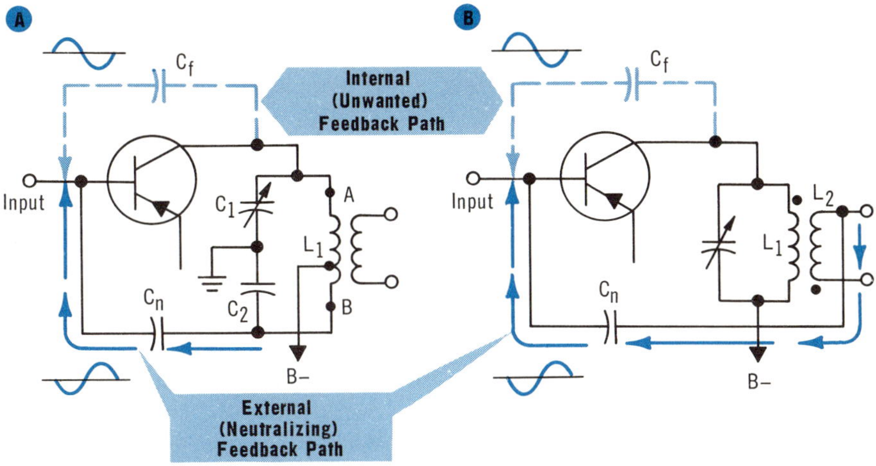

Neutralization prevents oscillation by feeding back to the input a signal that is 180° out of phase with the unwanted internal feedback signal. The two signals cancel each other enough to provide stable operation

For transistor r-f voltage amplifiers, neutralization is almost always used. The unwanted collector-to-base feedback, although not as likely to cause oscillation as is the case in tube circuits, causes undesirable distortion of the amplifier waveform. To cancel this feedback, various types of neutralizing circuits can be used.

In the circuit shown in A, the unwanted collector-to-base feedback is represented by the capacitor C_f. Coil L_1 and capacitor C_1 make up the output tuned circuit, and capacitor C_2 causes the signals at the opposite ends of L_1 (points A and B) to be of opposite polarity with respect to ground. Thus, since the unwanted feedback signal is coupled to the base from the top (point A) of L_1, and the neutralizing signal is coupled from the bottom (point B) through capacitor C_n, the two feedback signals are 180 degrees out of phase. The circuit components are selected so that the neutralizing signal has sufficient amplitude to cancel enough of the internal feedback to ensure stable operation.

Another method of neutralization is shown in B. Here the neutralizing signal is obtained from the secondary, L_2, of the coupling transformer. Because of the 180-degree phase reversal that takes place in a transformer, the signal at the top of the secondary (L_2) is 180 degrees out of phase with the signal at the top of the primary (L_1). In this way, the unwanted feedback signal coupled to the base through C_f is out of phase with the neutralizing signal coupled through C_n. When this method is used, care must be taken that the neutralizing signal is taken from the correct side of the coupling transformer secondary, or else the two feedback signals will reinforce instead of cancel each other.

the r-f gain control

Very often, the level, or amplitude, of the output signal from an r-f voltage amplifier must be capable of being *manually controlled.* This permits relatively constant-amplitude output signals to be obtained even though the different input signals to which the amplifier is tuned may vary widely in amplitude.

Manual control of the output level is provided by the r-f gain control. The gain control is practically always a *variable resistor,* which can accomplish its function in two ways, depending on where it is used in the circuit. In one of these ways, the gain control affects the *signal itself,* either by shunting a portion of the signal to ground, or otherwise attenuating the signal. This method is generally unsatisfactory because of the very low strength of r-f signals. The other way in which the gain control can work is to leave the signal itself alone and change the gain of the r-f amplifier. By reducing the amplifier gain, the amplitude of the output signal is correspondingly lowered, and vice versa.

Some commonly used methods by which the r-f gain control reduces amplifier gain are illustrated. In A, variable resistor R is the gain control. It can be located in the cathode circuit *or* the screen grid circuit. When it is used in the cathode, it varies the *cathode bias* of the stage. With R at its minimum setting, the cathode bias is at its lowest value, being supplied only by resistor R_1. The gain of the stage is, therefore, maximum. Increasing the resistance of R causes the bias to increase, with a resulting decrease in stage gain. This type of gain control is used with *remote-cutoff* (variable-mu) *tubes,* since varying the grid bias of these tubes causes a smooth change in gain.

When the gain control is placed in the screen grid circuits, the value of R determines the screen grid current and, therefore, the screen grid voltage. By varying R, the screen grid voltage and the gain of the stage are controlled. This, though, is not as effective as the previous gain control.

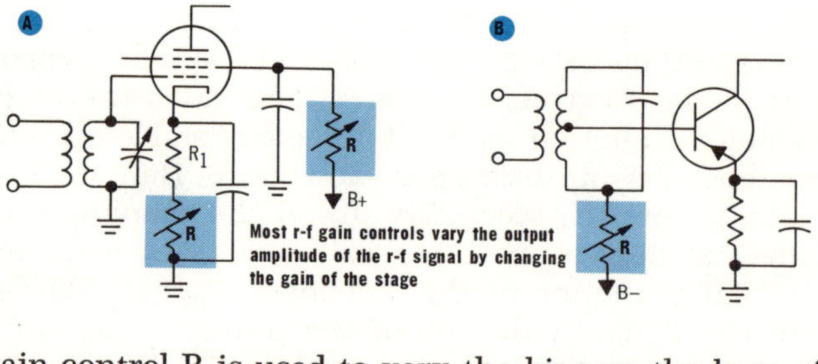

Most r-f gain controls vary the output amplitude of the r-f signal by changing the gain of the stage

In B, gain control R is used to vary the bias on the base of the stage. Increasing or decreasing the setting of R changes the base voltage and, therefore, the gain of the stage.

Frequently, the gain of r-f amplifiers is controlled automatically as well as, or instead of, manually. This is called *automatic gain control* (AGC), and is covered in Volume 7.

r-f traps

Sometimes, even with tuned circuits having the required bandwidth, unwanted large-amplitude signals are passed by the tuned circuits and are amplified along with the desired signal, causing interference and distortion. This occurs when the unwanted signals are *very strong* compared with the desired signal. Unwanted strong signals such as these can be removed from a receiver input before they can cause interference by circuits called *r-f traps,* or wave traps.

Basically, an r-f trap is a circuit that causes almost complete attenuation of the interfering signal without affecting any other signal frequency. Two common types of r-f traps are shown below. Since both of these traps are in the antenna circuit they eliminate the undesired signal before it can reach the r-f amplifier.

A **Parallel Resonant R-F Trap**

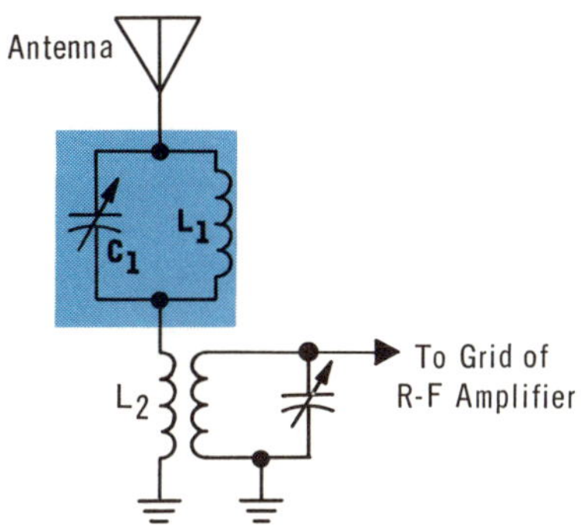

B **Series Resonant R-F Trap**

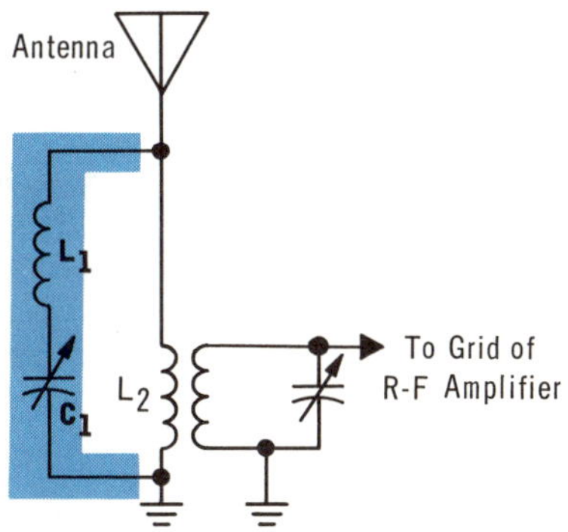

R-f traps provide almost complete attenuation of an undesired signal frequency with little effect on other frequencies

In A, the trap consists of a parallel resonant circuit in series with the antenna and the antenna coil. The trap is tuned by capacitor C_1 to the frequency of the undesired signal. It thus offers a very high impedance at this frequency, preventing the signal from being developed across antenna coil L_2 and being coupled from there to the input of the r-f amplifier. To all other frequencies, though, the trap offers a low impedance, so appreciable currents of all but the unwanted frequency can flow in the antenna circuit.

The trap shown in B is a series resonant circuit that shunts the antenna coil. It is tuned to the frequency of the interfering signal, so it has minimum impedance at this frequency and a high impedance at all other frequencies. It thus shunts the unwanted signal currents around the antenna coil, but allows currents of all other frequencies to flow through L_2 and develop a signal voltage across it. Quite often, combinations of series and shunt traps are used.

bandswitching

In many equipments, an r-f amplifier must be capable of operating over a *very wide frequency range.* This means that the tuned circuits must be able to be tuned to any frequency within this wide range. Furthermore, the tuned circuits have to maintain the desired bandwidth regardless of the frequency to which they are tuned.

Usually, only one component (either L or C) of a tuned circuit is *variable.* When the circuit must be tunable over a very wide range, the variable component has to be adjustable over an almost impractical range of values. This difficulty could be avoided by making both components of the tuned circuit variable. However, when more than one tuned circuit is used, this would cause significant problems in tuning all of the circuits to the same frequency. Instead, what is usually done is to make one component variable and *switch* different fixed values of the other component into the circuit for different portions of the overall tuning range.

Bandswitching makes use of one variable component and any one of multiple fixed components

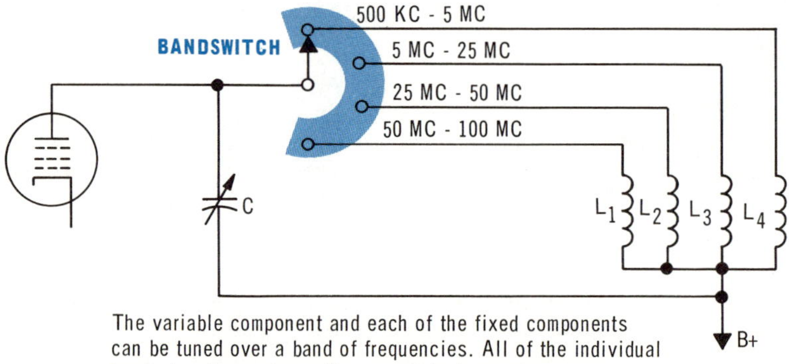

The variable component and each of the fixed components can be tuned over a band of frequencies. All of the individual bands together make up the overall tuning range

For example, assume the case of a parallel resonant circuit in the output of any r-f amplifier that operates over the frequency range of 500 kilocycles to 100 megacycles. A variable capacitor could be used for the capacitive circuit element, and four separate coils for the inductive element. The capacitor and all four coils are connected to a switch, called a *bandswitch,* which allows either one of the coils to be connected in parallel with the capacitor. Thus, if the overall tuning range is divided into subranges, or bands, of 500 kc–5 mc, 5 mc–25 mc, 25 mc–50 mc, and 50 mc–100 mc, the appropriate coil is switched into the circuit for each range.

When bandswitching is used, the variable component, say the capacitor, only has to have a limited range. This is because the same values of the variable capacitor produce different resonant frequencies when combined with the values of the different coils.

cascade r-f voltage amplifiers

R-f voltage amplifiers perform a variety of functions, depending on the type of equipment they are used in. In all cases, though, two of their most important functions are to produce signal amplification, or *gain,* and to have a good degree of frequency discrimination, or *selectivity.*

Both of these characteristics are directly related to the *number* of r-f amplifier stages used and can be improved or increased by connecting a number of amplifier stages in *cascade.* Essentially, this means that similar-type amplifiers are used in series, with the output of one connected to the input of the next, and so on. For practical purposes, the total gain of cascaded amplifiers can be considered to be the *product* of the gains of the individual stages. Thus, if three stages are used, each having a gain of 50, the overall gain is 50 × 50 × 50, or 125,000.

You should understand here that although gain is increased by cascading individual amplifiers, it does not mean that any extremely weak signal can be amplified to a usable level merely by adding stages. Noise received at the antenna along with the signal, as well as noise generated by the amplifier stages themselves, is amplified together with the signal. If this noise level is such that the *signal-to-noise ratio* at the output of the last amplifier stage is below a certain level, the signal will be unusable regardless of how much it has been amplified.

Cascading individual stages improves selectivity by increasing the number of tuned circuits. As shown, increasing the number of tuned circuits does not greatly affect the peak of the amplifier response curve, but significantly steepens the sides of the curve. This means that the response is much *sharper,* with the signal output decreasing greatly for small changes in frequency above and below resonance.

Connecting r-f amplifiers in cascade increases overall gain and sensitivity

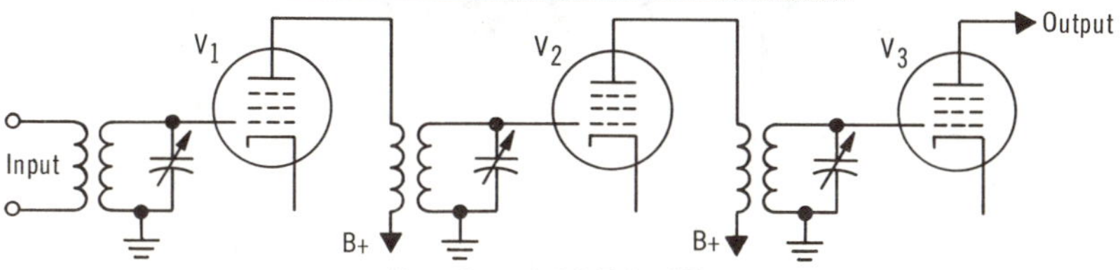

Three Cascaded R-F Amplifiers

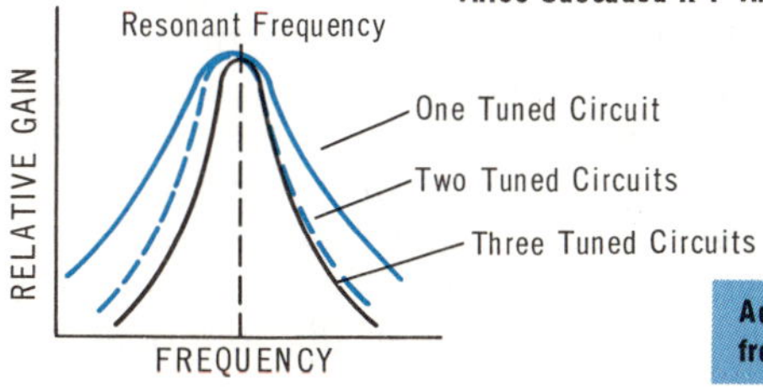

Additional tuned circuits sharpen the frequency response curve of an r-f amplifier

ganged tuning

A TWO-SECTION GANGED TUNING CAPACITOR

Although cascading r-f amplifiers improves overall gain and selectivity, it also increases the number of tuned circuits that must be tuned to the same frequency. Each time a new signal is to be received, every tuned circuit must be *simultaneously* retuned to the new signal frequency.

This simultaneous tuning is generally accomplished by having all of the tuned-circuit capacitors mounted on a *common shaft.* When the shaft is rotated, the plates of each capacitor move into or out of mesh by the same amount, so the capacitance of each is changed equally. Such a capacitor is called a *ganged tuning capacitor* and has as many individual sections as there are tuned circuits. In effect, then, the sections of the ganged tuning capacitor make up individual but identical capacitors whose values are changed equally by rotating the common shaft. When the shaft is rotated, the capacitance value, and, therefore, the resonant frequency of each tuned circuit is changed equally.

On schematics, ganged capacitors are connected by dashed lines to indicate that they are tuned simultaneously. In effect, the dashed lines represent the common shaft

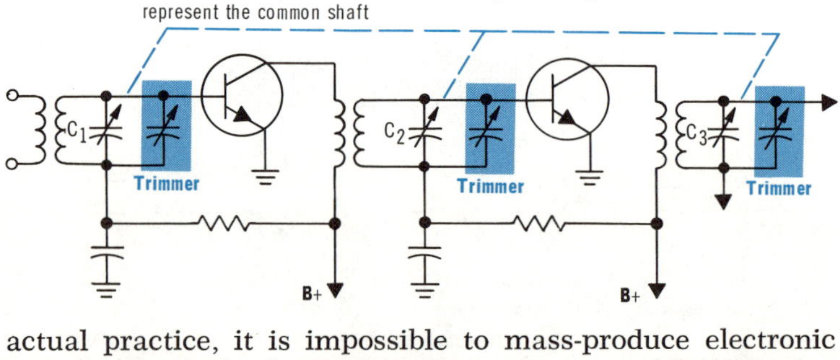

In actual practice, it is impossible to mass-produce electronic components with identical values. Thus, variations caused by manufacturing tolerances exist between the supposedly identical sections of a ganged capacitor, as well as between the coils used in the tuned circuits. To compensate for these differences, every section of a ganged capacitor has a small capacitor called a *trimmer capacitor* in parallel with it. The trimmer capacitors are adjusted so that each section of the ganged capacitor has the required value to provide essentially identical tuning of the tuned circuits.

cascode r-f amplifiers

A somewhat special but very common type of r-f amplifier is called a *cascode* r-f amplifier. A tube-type cascode circuit consists of a *grounded-cathode* stage driving a *grounded-grid* stage, while in a transistor cascode circuit, a *grounded-emitter* stage drives a *grounded-base* stage. This could be done with capacitor coupling, but a cascode circuit usually employs *direct coupling* between stages, with the signal coupled directly from the plate or collector of the first stage to the cathode or emitter of the second stage. In effect, the two stages are connected in *series*, with the input circuit of the second stage acting as the output load for the first.

When tubes are connected in cascode, triodes can often be used without neutralization. This is because both tubes form a voltage divider across the B+ supply, so the plate voltage and, therefore, the gain of the first stage are relatively low. As a result, the feedback voltage through the plate-to-grid interelectrode capacitance is not, in many cases, sufficient to cause oscillation. As the second stage is connected in the grounded-grid configuration, it does not require neutralization under any circumstances. By using triodes, therefore, a cascode r-f amplifier has the low-noise characteristics inherent in triode amplifiers. At the same time, the total gain of the two stages is about the same as would be obtained with a single pentode stage.

When transistors are connected as a cascode r-f amplifier, neutralization is also generally not needed. This is because of the stabilizing effect that the very low input impedance of the grounded-base stage has on the grounded-emitter stage.

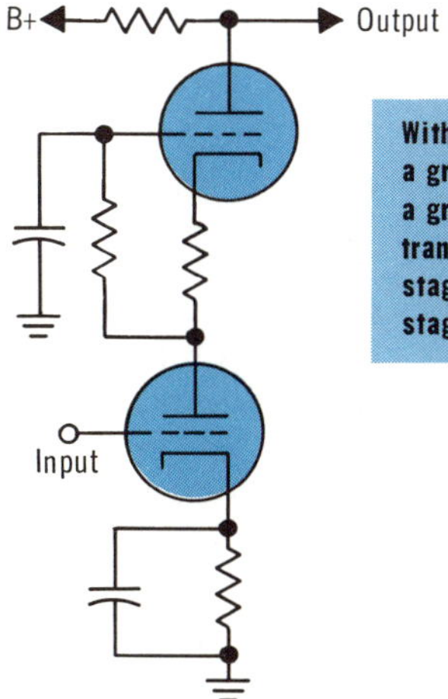

With tubes, a cascode circuit is a grounded-cathode stage driving a grounded-grid stage. With transistors, a grounded-emitter stage drives a grounded-base stage

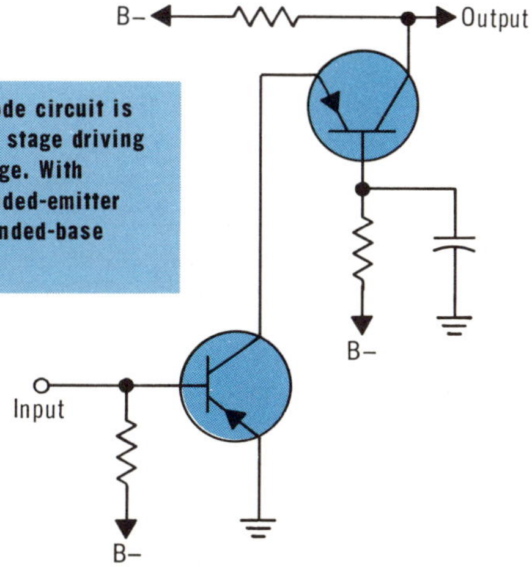

In the tube version, the grid of the output stage is at ground potential for the r-f signal. In d-c terms, though, the grid is returned to the plate of the first stage. If this were not done, the high voltage on the cathode of the output stage would cut the stage off

circuits for cascode r-f amplifiers

A typical, tube-type cascode r-f amplifier is shown in A. V_1 is a grounded-cathode stage, and V_2 is a grounded-grid stage. The input signal is coupled through transformer T_1 to the series-tuned circuit consisting of coil L_1 and variable capacitor C_1. The signal is developed across C_1 and applied to the grid of V_1. Stage V_1 operates class A, with cathode bias supplied by resistor R_1 and capacitor C_2.

A Tube-Type Directly-Coupled Cascode R-F Amplifier

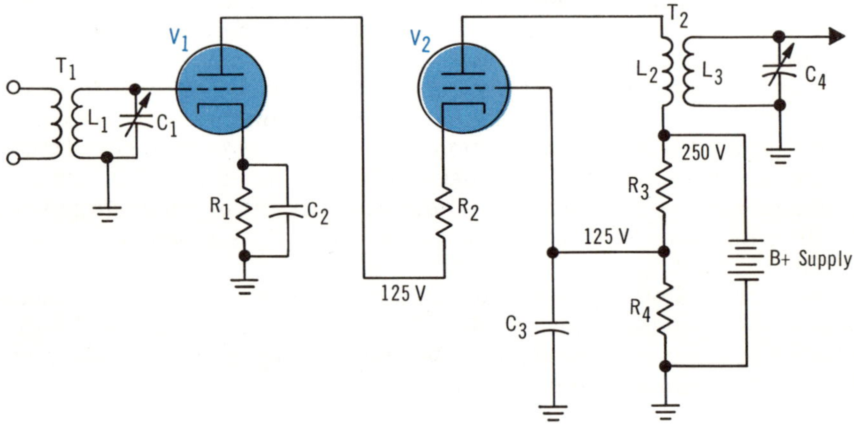

The plate current of V_1 flows to the B+ supply through stage V_2. This means that the *entire* output current of V_1, both the r-f component and the d-c component, flows through R_2, which functions as the cathode resistor of stage V_2. The r-f component of the voltage drop across R_2 is the input to the stage, while the d-c component of the voltage drop, which is 125 volts in this circuit, creates a large cathode voltage. If the grid of the stage were connected *directly* to ground, a large bias would result and cut off the stage. To prevent this, the grid is kept at r-f ground potential by capacitor C_3, and is returned at the same time to a 125-volt tap on the B+ supply. The bias for the second stage is then dropped across R_2, and so is the signal voltage.

Since the cathode fluctuates with the signal voltage, and the grid voltage is fixed, the signal is then amplified in tube V_2. The amplified output of stage V_2 is developed across coil L_2 and transformer-coupled to a tuned circuit serving as the input to a following stage. The cascode amplifier is tuned to the desired frequency by ganged capacitors C_1 and C_4. The trimmers for these capacitors are not shown.

circuits for cascode r-f amplifiers (cont.)

B Transistor-Type Directly-Coupled Cascode R-F Amplifier

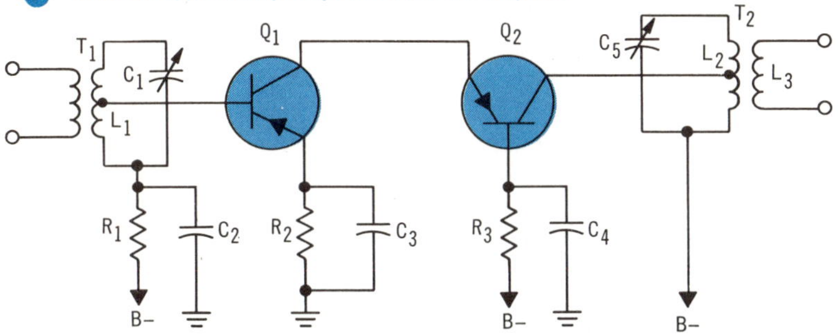

C Capacitor-Coupled Cascode R-F Amplifier

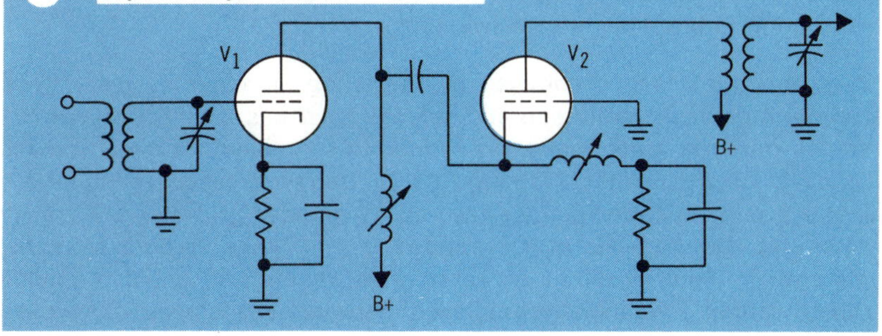

In the transistor cascode amplifier shown in B, transistor Q_1 is connected as a grounded-emitter stage, and Q_2 is connected as a grounded-base stage. The input to the base of Q_1 is from the tuned circuit consisting of tapped coil L_1 and capacitor C_1. Capacitor C_2 bypasses the r-f signal around bias resistor R_1 so that the maximum input signal is developed across the tuned circuit. Temperature and bias stability for Q_1 are provided by resistor R_2 and capacitor C_3. The emitter-collector circuit of Q_1 and the base-emitter circuit of Q_2 form a series circuit, providing Q_1 with excellent stability. Capacitor C_4 keeps the base of Q_2 at r-f ground potential to avoid degeneration. The output signal from Q_2 is developed across a tapped tuned circuit and transformer-coupled to the succeeding stage.

Diagram C shows how the cascode amplifier can be *capacitor-coupled.* It works similarly to ordinary cascoded stages, except that the output of V_1, which is a grounded-cathode stage, is coupled to the cathode of V_2, which is a grounded-grid stage. You can see that since the two stages are isolated from one another, they both can use the same B+ voltage.

summary

□ The bandwidth of an r-f amplifier is the range between the two frequencies at which the amplified output voltage falls to a value 0.707 times its value at the frequency to which the amplifier is tuned. □ Amplifiers with a narrow bandwidth have good selectivity; those with a wide bandwidth have poor selectivity. □ For single-tuned amplifier circuits, bandwidth is equal to f_R/Q. □ In various r-f amplifier circuits, such as a triode grounded-cathode stage, positive feedback occurs, which can cause the stage to oscillate. To prevent oscillation, an external signal is fed back to the input to cancel the unwanted feedback. This is called neutralization. □ Pentodes or grounded-grid amplifiers do not require neutralization.

□ A variable resistor in either the cathode or screen circuit can be used to provide r-f gain control. □ R-f traps, or wave traps, remove unwanted strong signals before they reach the r-f amplifier. □ When an r-f amplifier must be tunable over a wide range, the use of a single tuned circuit is impractical. In this case, one component is made variable, and different fixed values of the other component are switched into the circuit for different portions of the overall tuning range. This is bandswitching.

□ Very high gain can be obtained by cascading amplifier stages. □ While cascading improves overall gain and selectivity, it has a disadvantage in that all stages must be gang tuned. To reduce this problem, trimmer capacitors are used to ensure identical tuning. □ A tube-type cascode amplifier consists of a grounded-cathode stage feeding a grounded-grid stage. A transistor cascode amplifier consists of a grounded-emitter stage feeding a grounded-base stage. □ Neutralization is generally not required in cascode amplifiers. □ Direct coupling is usually employed between the stages of a cascode amplifier.

review questions

1. How do cascaded and cascoded amplifiers differ?
2. Why is neutralization required for triode r-f amplifiers?
3. What are r-f traps and why are they used?
4. What is meant by the *signal-to-noise ratio*?
5. If the gain of each stage of a three-stage cascaded amplifier is 10, what is the overall gain?
6. A gain of 9000, minimum, is required. How many cascaded stages are required if each stage has a gain of 20?
7. Draw an r-f amplifier with a gain control.
8. What is meant by the *selectivity* of an amplifier?
9. Why are trimmer capacitors used in cascaded r-f amplifiers?
10. What is meant by *bandswitching*?

the r-f buffer amplifier

Although r-f buffer amplifiers are used in many kinds of electronic equipment, it is in high-power radio transmitters that their use is most extensive. As you know, in a transmitter, a *low-level* r-f signal is produced by an oscillator. This low-level signal is usually then increased in frequency and power level by succeeding stages. If the oscillator in a transmitter was to feed the output to a stage that required a large amount of *input driving power*, such as an intermediate or power amplifier, the oscillator would be loaded down and would *shift* from its natural frequency. This is especially true if the stage fed by the oscillator was keyed. Each time the transmitter was keyed, the load on the oscillator would change and its output frequency would shift. To prevent this loading of the oscillator, a buffer amplifier is placed in the circuit *directly after* the oscillator.

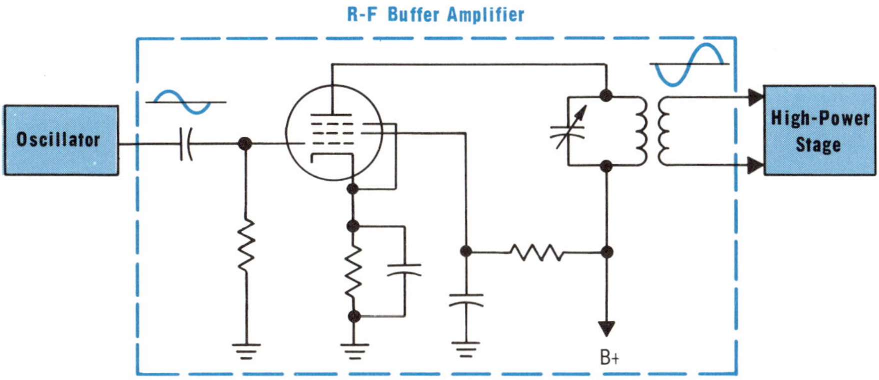

Ideally, an r-f buffer amplifier is a class A voltage amplifier. It produces no loading on the oscillator, and delivers as an output a voltage-amplified version of its input signal. Its purpose is to isolate stages

Ideally, the buffer amplifier is a *class A* r-f *voltage* amplifier. The high-input impedance of such a stage has no loading effect on the oscillator, and the amplified output contains sufficient power to drive the following stage easily. Very often, though, the luxury of an ideal buffer amplifier cannot be afforded, so less efficient but more practical circuits are used. For example, the buffer might be operated as a low-level class B or C power amplifier. It would produce some loading of the oscillator, but would be performing the added function of power amplification at the same time. As another example, the buffer might also be used as a frequency multiplier. Again, it would be operated class B or C, and would produce some loading on the oscillator.

You can see from the above that buffer amplifiers are really not a separate type of amplifier as far as *circuit configuration* is concerned. Many different types of circuits can perform the buffering function, and so, strictly speaking, are buffer amplifiers.

the frequency-multiplier stage

A frequency-multiplier stage receives as its input a sinusoidal signal, usually from an oscillator or buffer amplifier, and delivers as its output a sinusoidal signal from a tank circuit having a frequency that is some *multiple* of the input signal frequency. If the output has *twice* the frequency of the input, the stage is called a *frequency doubler;* if the output is *three times* the input, it is a *frequency tripler;* and so on.

Frequency multipliers are used extensively in AM radio transmitters that operate in the VHF region and higher. At these high frequencies, it is difficult to design oscillators that are stable. With frequency multipliers, oscillators can be used that have frequencies many times less than the transmitter output frequency. For example, if an oscillator having a frequency of 4 megacycles was used with two triplers and one doubler, its frequency would be multiplied eighteen times to a final frequency of 72 megacycles.

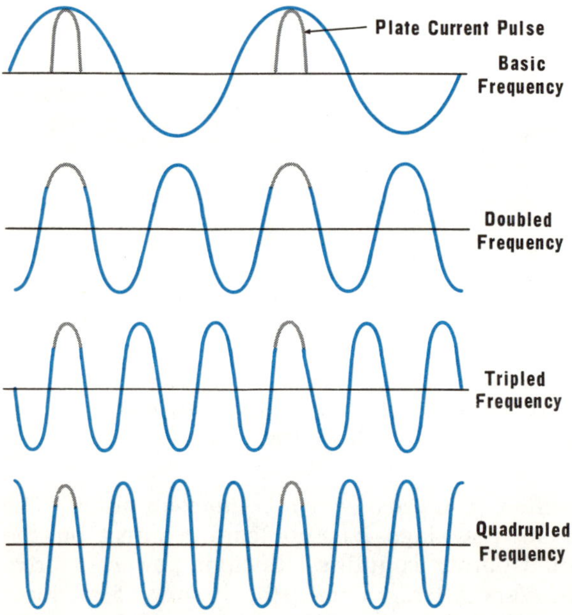

In frequency multipliers, the plate current pulses for the basic frequency reinforce the tank circuit oscillations. In a doubler, they reinforce the peak of every other sine wave; in a tripler, they aid the peak of every third sine wave; in a quadrupler, they aid every fourth sine wave; and so on

Frequency multipliers are also widely used in FM transmitters. When the FM carrier is shifted in frequency by the modulation, the deviation is quite small. However, by sending the signal through a frequency multiplier, both the center frequency and the deviation are increased. For example, a center frequency of 10 megacycles that shifts to 10.001 megacycles, for a shift of 1 kilocycle, can be converted to a center frequency of 90 megacycles with a shift to 90.009 megacycles by being sent through two tripler stages. The shift is thus increased nine times, from 1 kilocycle to 9 kilocycles.

the frequency-multiplier stage (cont.)

Frequency Doubler Waveforms

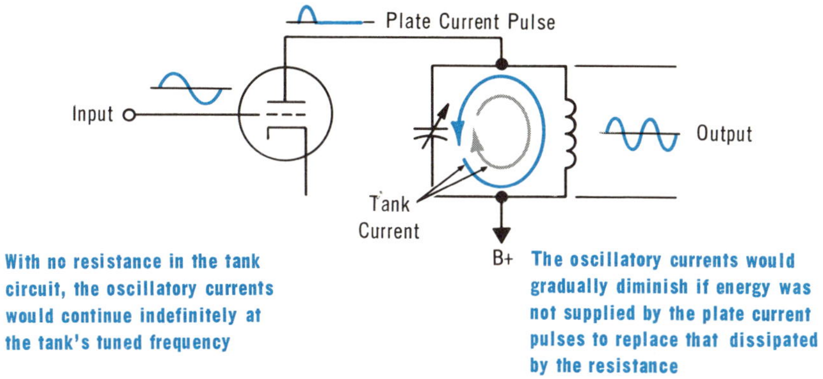

Frequency multipliers are essentially class C r-f amplifiers that have a parallel resonant circuit, or tank circuit, as their plate load. Plate current flows only during the positive peaks of the input signal, but because of the oscillatory or flywheel current produced in the tank circuit, a *sinusoidal output* is produced. The frequency of the output signal is determined by the resonant frequency of the tank circuit.

In theory, oscillatory currents can be set up in the tank circuit, and a sinusoidal output produced, at any multiple (or harmonic) of the circuit input frequency. In practice, though, the oscillatory currents, and, therefore, the amplitude of the output signal, decrease for each increase in harmonic. For example, if the tank circuit is tuned to the second harmonic of the input frequency, the output is 65 percent of what it would be if the tank was tuned to the input frequency. If the tank is tuned to the third harmonic, the output is 40 percent of what it would be if it was tuned to the input frequency. Because of this reduction in output, a frequency-multiplier stage is rarely tuned beyond the *fifth* harmonic of its input frequency.

You should understand here that if the tank circuit of a multiplier is tuned to any frequency other than a multiple of the input, oscillatory currents cannot flow in the tank. The output of the stage is therefore not sinusoidal.

In tube circuits, triodes with their high plate efficiency can be used for frequency multipliers without the necessity for neutralization. This is because the grid and plate signals have *different frequencies*.

A one-tube multiplier circuit is shown in A on the next page. The r-f input signal is applied to the grid of the stage through capacitor C_1 and developed across coil L_1, which has a large reactance at the r-f frequency. The stage is biased well beyond cutoff by the fixed bias battery, so the tube conducts only on the positive peaks of the input signal. The resulting pulses of plate current are rich in harmonics, since they are a distorted version of the input signal.

frequency-multiplier circuits

The frequency of the plate tank circuit is set to the desired harmonic by setting switch S for the desired tank coil and then fine-tuning capacitor C_2. The output of this type of circuit is usually RL or RC coupled to the following circuit, which may be another frequency multiplier or some type of power amplifier. Although fixed bias is shown for the circuit, a combination of grid-leak and cathode bias is often used.

A circuit that can be used for the efficient production of even-order harmonics in its output is the *push-push multiplier* shown in B. When used as a doubler or quadrupler, this circuit has a greater power output and higher plate efficiency than a single-tube multiplier. As shown, the *grids* of a push-push stage are connected in *push-pull*, and the *plates* are connected in *parallel*. As a result, the signals applied to the grids of the two tubes are 180 degrees out of phase, so that a pulse of plate current flows in one tube on the positive peaks of the input signal, and a pulse of plate current flows in the other tube on the negative peaks of the input signal. Two pulses of current are therefore delivered to the tank circuit during each full cycle of the input signal, instead of only one pulse as in the one-tube multiplier.

If both the inputs and outputs of two tubes are connected in push-pull, the circuit can be used as an efficient tripler, since such a circuit produces only odd-harmonic frequency multiplication.

A One-Tube Multiplier

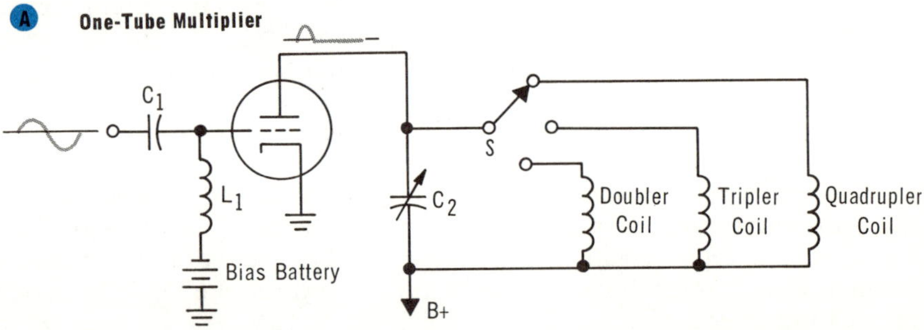

B Push-Push Multiplier (Used as Doubler)

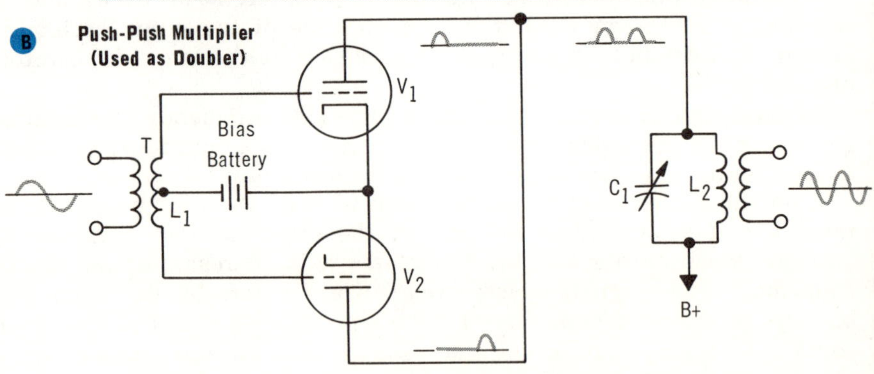

r-f power amplifiers

R-f power amplifiers are used to increase the power level of an r-f signal applied to their input. Their principal application is in electronic transmitters, where they boost the r-f carrier signal to the high power levels required to drive transmitting antennas. R-f power amplifiers can be classified in various ways: by their relative output levels, their class of operation, the type of tubes used, or their circuit configuration. In practically all cases, though, they require relatively large input driving power and use tuned circuits as their output load.

R-f power amplifiers operate either class B or C, which means that plate current flows for only a *portion* of the input signal cycle. The output is *sinusoidal,* nevertheless, because of the flywheel action of the tuned circuit in the output. Since *grid current* flows in a class B or C amplifier when the input signal drives the grid positive, grid-leak bias is used extensively in r-f power amplifiers. However, a certain amount of fixed or cathode bias is usually also provided, inasmuch as the grid-leak bias is lost when there is no input to the stage; for example, when the transmitter is keyed. Without this additional bias, excessive current would flow through the tube when the grid-leak bias was lost and damage would result.

Some of the various types of power amplifier circuits are shown.

Grounded-Cathode, Triode R-F Power Amplifier

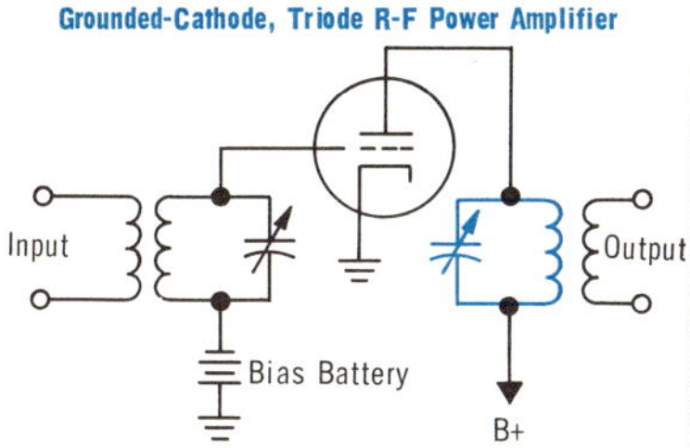

Grounded-Grid, Triode R-F Power Amplifier

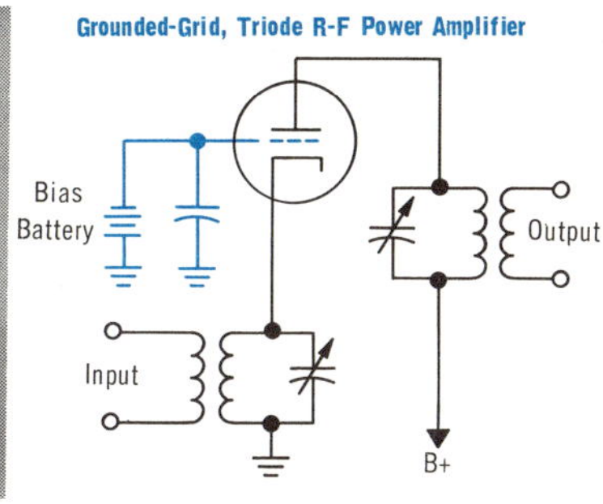

The grounded-cathode power amplifier circuit is essentially the same as a frequency multiplier circuit. The basic difference is that the output tank circuit is tuned to the input frequency rather than to a harmonic of that frequency. Although not shown, this circuit must be neutralized

Push-Pull R-F Amplifier Using Tetrodes

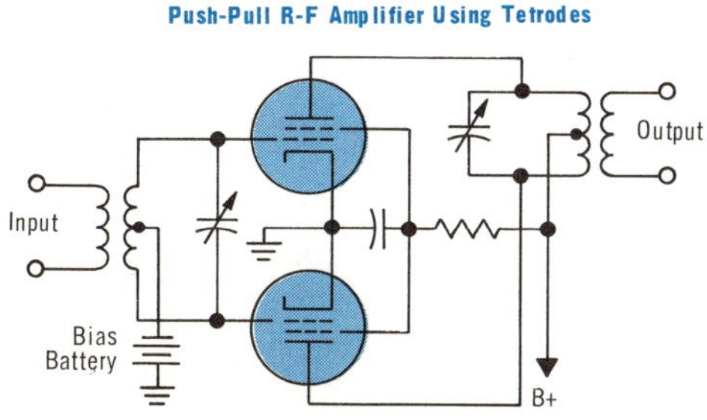

In the grounded-grid circuit, the grid acts as a shield between the input and output circuits, preventing feedback. Therefore, a triode can be used without the need for neutralization

tube vs. transistor r-f power amplifiers

Although both tubes and transistors can be used as r-f power amplifiers, they are significantly different in their electrical capabilities. Tubes have been used for many years, and it is a relatively easy matter to design and construct tube-type r-f amplifiers that can operate at VHF, UHF, and higher frequencies and deliver hundreds and even thousands of watts of output power.

Transistors, on the other hand, still have *frequency* and *power* limitations that make their use as r-f power amplifiers fall far short of that of tubes. Special transistors have been developed that can deliver a few hundred watts of output power at useful r-f frequencies; but in most cases, the cost of the transistor is much greater than that of a comparable tube circuit. Consequently, transistor r-f amplifiers are for the most part limited to special applications involving low power requirements, such as walkie-talkie transmitters, or applications where the greatly-decreased size and weight of transistor circuits are more important than the additional cost. An example of this would be the transmitters used in satellite telemetry systems.

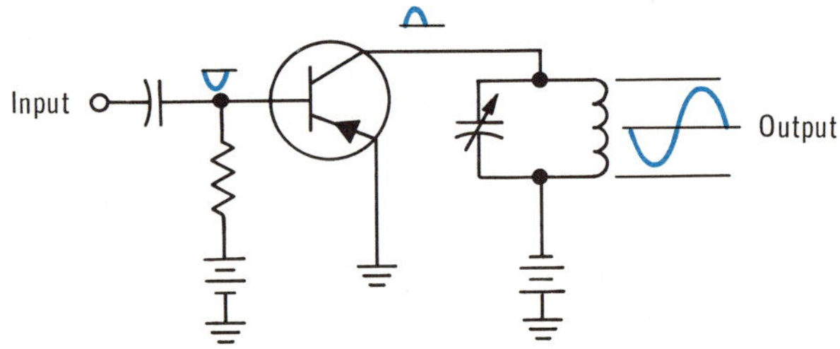

Although the base-emitter circuit is normally biased in the forward direction, for class C operation it is reverse-biased. This is necessary to bias the stage well beyond cutoff

If a transistor is used as an r-f power amplifier, it is usually operated class C. For most of the input cycle, therefore, only a small amount of "cutoff" current flows in the collector circuit. For the grounded-emitter circuit shown, this means that the transistor only conducts in bursts, and these bursts occur during the positive peaks of the input signal. But, as in tube circuits, the bursts of collector current are converted into a sinusoidal output by the action of the tank circuit in the collector output. A unique feature of the class C transistor amplifier is that in order to achieve a bias point considerably beyond cutoff, the emitter must be biased the *reverse* of what is considered normal. Thus, in the circuit shown, the emitter is biased positive with respect to the base.

effect of modulation

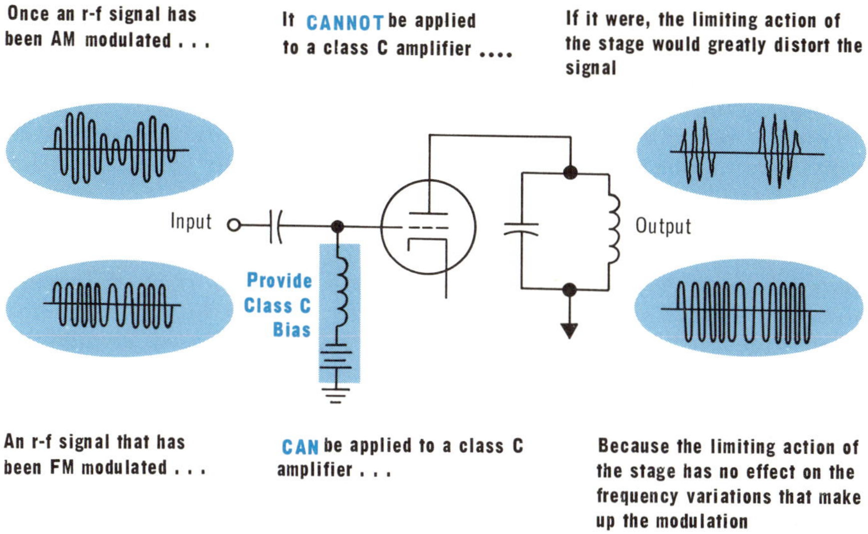

In many transmitters, a series of cascade r-f power amplifiers is required to produce the necessary output power. The last stage is called the *final* power amplifier, or merely the power amplifier, and the preceding stages are called *intermediate* amplifiers, or *drivers*. If the r-f signal is to be *modulated* (either AM or FM), the modulation can be introduced at various places in the transmitter.

For AM modulation, the point at which the modulation occurs has an important effect on the type of r-f power amplifier circuits that are used. This is because the modulation varies the instantaneous peak-to-peak *amplitude* of the r-f signal. These variations, which make up the modulation envelope, must be preserved once modulation takes place. Any stage that amplifies the signal after it has been modulated has to be *linear*, which means class A, or class B push-pull. The *unmodulated* r-f signal, though, can be amplified by a class C stage. Since class C amplifiers have the highest efficiency, high-power AM transmitters are most efficient when the modulation takes place at the output of the final power amplifier. All of the stages, including the last, can then be operated class C.

In FM transmitters, modulation takes place *before* any of the power amplifier stages. Every power amplifier therefore amplifies the modulated signal. However, the FM signal has no envelope that can be distorted by the characteristics of a class C stage. The modulation is merely a shift in the frequency of a constant amplitude r-f signal, and these frequency variations can be reproduced in a class C amplifier without distortion. Class C amplifiers, with their high efficiencies, are therefore generally used for the power stages of FM transmitters.

AM vs. FM r-f power amplifiers

You have already seen one of the differences between r-f power amplifiers used for AM signals and those used for FM. In addition, there is one other significant difference. This is the Q or *selectivity* of the tuned input and output circuits.

Tuned circuits used with power amplifiers in AM transmitters generally have higher Q's and, therefore, greater selectivity. The reason for this is that in AM transmitters the tuned circuits usually only process the unmodulated r-f signal. There are no *side bands,* so the circuits can be *sharply* tuned to the r-f frequency. In FM transmitters, on the other hand, the r-f power amplifiers and their tuned circuits process the modulated signal. This signal contains side bands both above and below the center frequency, and these side bands must be passed *equally* by the tuned circuits if the signal is not to be distorted. The response of these tuned circuits, therefore, must be broad enough to accommodate all of the side bands.

FOR AM

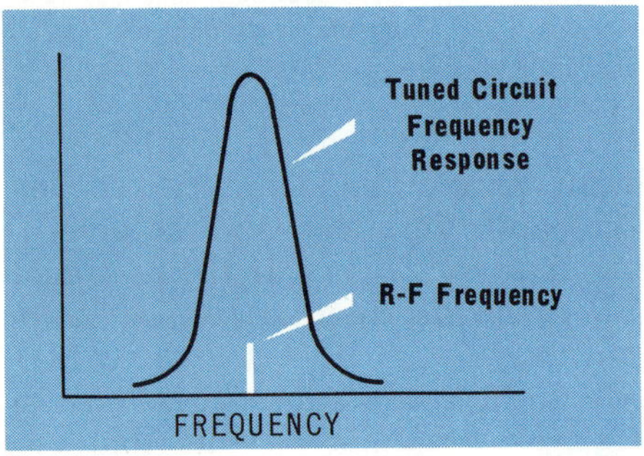

When only a single frequency, such as the unmodulated AM carrier, must be passed, a tuned circuit should have a high Q, and, therefore, good selectivity

FOR FM

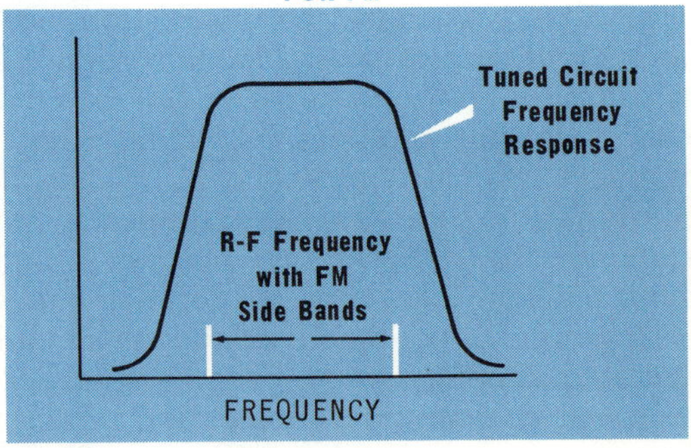

When a signal containing side bands must be passed, a tuned circuit requires a low enough Q so that all the side bands will be passed equally

The need for lower Q's in the r-f amplifier tuned circuits of FM transmitters has the effect of making triodes, in general, less desirable for FM than for AM power amplifiers. This is because of the relatively large *output capacitance* of triodes. The shunting effect of this capacitance on the plate tank circuit tends to increase the Q of the circuit, resulting in too narrow a bandpass.

input circuits

Since an r-f tube-type power amplifier is operated class C or class B push-pull, its input impedance is not extremely high like that of a class A amplifier. For efficient power transfer, therefore, there must be an impedance match between the grid input impedance of the r-f power amplifier and the plate output impedance of the previous stage. There are many circuit arrangements that can be used to provide this impedance match, with the one used chosen on the basis of the particular requirements of the application. Some of these circuit arrangements are illustrated here.

In this capacitance-coupled arrangement, L and C make up the output tank circuit of V_1. Coil L is tapped to step up the signal voltage from V_1 and present the power amplifier input circuit with the proper impedance

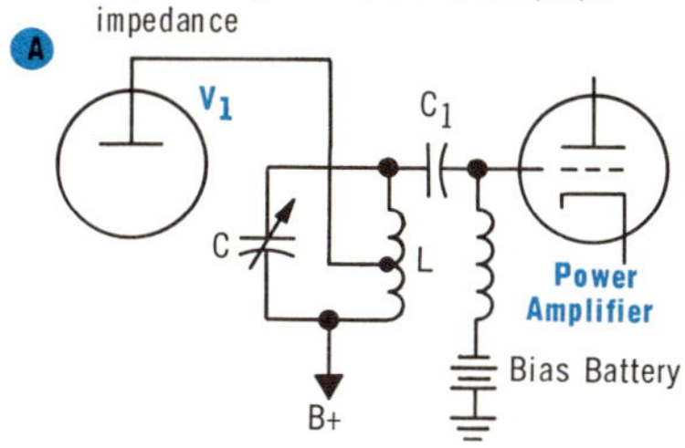

The signal is taken from the top of the tank and coupled to the grid of the power amplifier through C_1. This circuit is used to match a lower output impedance to a higher input impedance

This capacitance-coupled arrangement is similar to A, except that the tuned circuit is in the grid circuit of the power amplifier

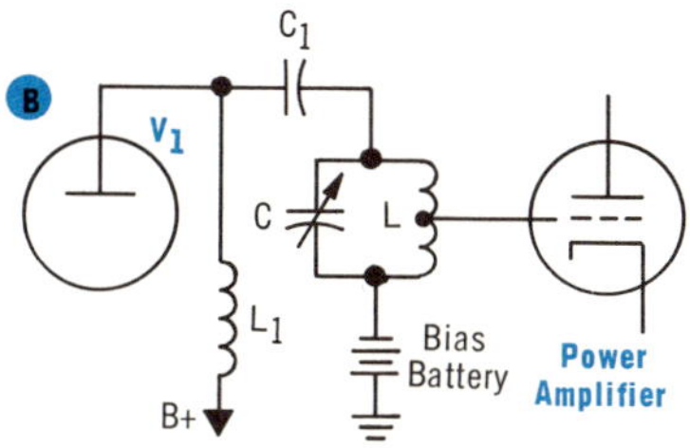

The tapped tank coil steps the signal voltage down and presents a lower impedance to the input of the power amplifier. This circuit is used to match a higher output impedance to a lower input impedance

This capacitance-coupled arrangement is somewhat similar to A. However, by applying plate voltage for V_1 through the tap of the coil, the voltage at the bottom of the tank circuit has the correct phase for neutralization

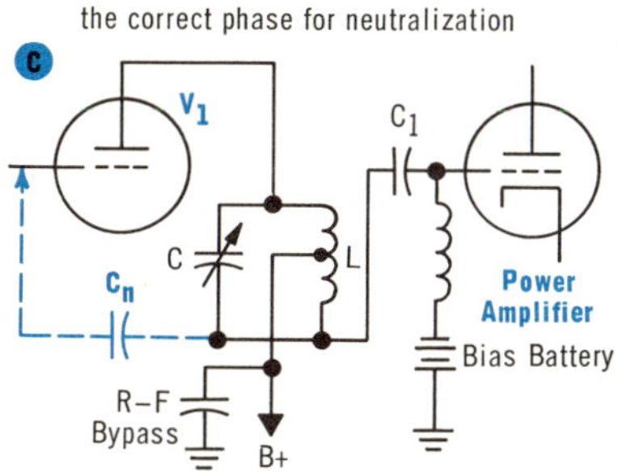

A triode can therefore be used for V_1 and a neutralizing voltage fed back to the grid of V_1 through capacitor C_n. This is not possible with the circuits of A or B. Neutralization is covered in more detail later

In this arrangement, the signal is link-coupled from a tuned output circuit to a tuned input circuit. Essentially, the link is just a few turns of wire on either end joined by a length of transmission line

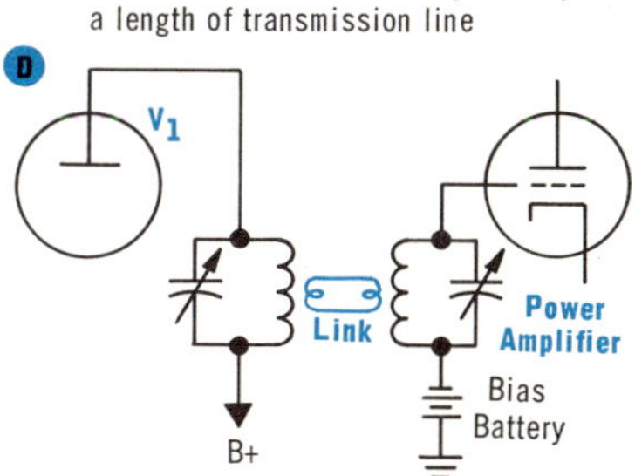

The link has a very small inductance, so the impedance of the coupling circuit is low, and power losses are minimized. An advantage of link coupling is that it allows the power amplifier to be located some distance from its driving stage

output circuits

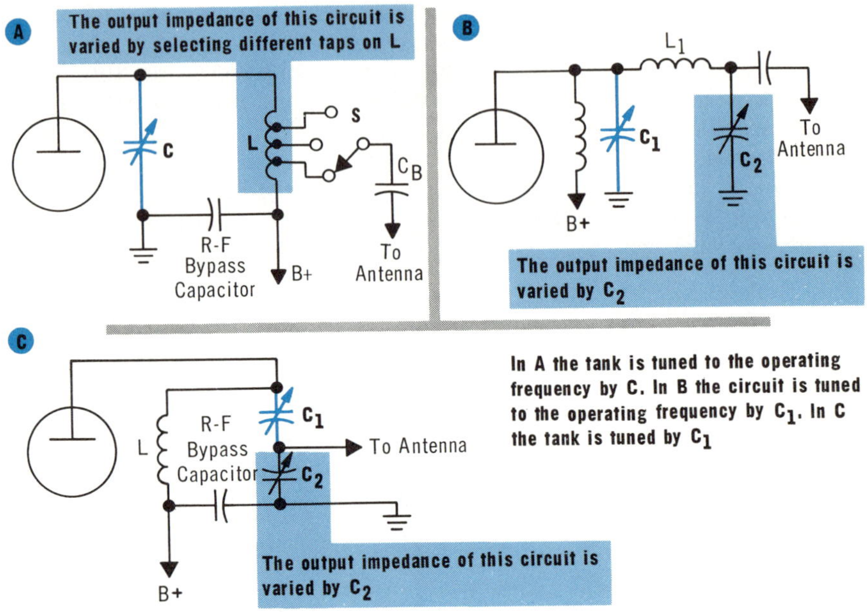

In A the tank is tuned to the operating frequency by C. In B the circuit is tuned to the operating frequency by C_1. In C the tank is tuned by C_1

When an antenna is fed directly from the r-f amplifier output tank circuit, which is commonly done in small portable transmitters, impedance matching is the main requirement. The tank circuit in the amplifier output must match the output impedance of the amplifier *directly* to the antenna impedance, which can vary over a wide range during operation. Therefore, a tank circuit must be used that can compensate for a wide range of impedance. Three basic circuits that can do this are shown.

In A, the coil of the output tank circuit has various taps, any of which can be selected by switch S. The bottom of the tank circuit is at ground potential for the r-f signal, so the bottom tap on the coil has a very low impedance, and each higher tap has a greater impedance. The desired impedance is selected by switch S, and the output is coupled to the antenna through capacitor C_B, which prevents the B+ supply voltage from reaching the antenna.

In B, a commonly-used output circuit, called a pi-network, is shown. Effectively, capacitor C_1 and C_2, with inductor L_1, make up the output tank circuit. L_1 and C_2 form an impedance-matching voltage divider with the relative voltage drops determined by the setting of C_2. C_1 is also variable and is used to tune the circuit to resonance at the operating frequency.

The circuit shown in C is a variation of the pi-network in which one of the capacitors is not grounded. The two capacitors (C_1 and C_2) make up the impedance-matching voltage divider, with the output taken from across C_2.

the antenna tuning unit

Quite frequently, a separate piece of equipment, called an *antenna tuning unit,* is used between the final power amplifier of a transmitter and the antenna. The antenna tuning unit performs two functions: (1) it provides a means of making the antenna *resonant* at any frequency over a relatively large range, thus making the antenna *purely resistive* at the operating frequency; and (2) it also provides a means of *matching* the antenna resistance to the output impedance of the r-f power amplifier.

Antenna circuits are covered in Volume 7. Now, though, you should understand that when an antenna is purely resistive at its operating frequency, maximum power transfer and radiation are achieved. By means of an antenna tuning unit, a transmitter can efficiently drive any one of many different antenna types having a wide variety of electrical characteristics.

Switches S_1 and S_2 are ganged so that various reactive components may be applied in series with the antenna. Coils L_2, L_3, and L_4 are variable to allow tuning over a range

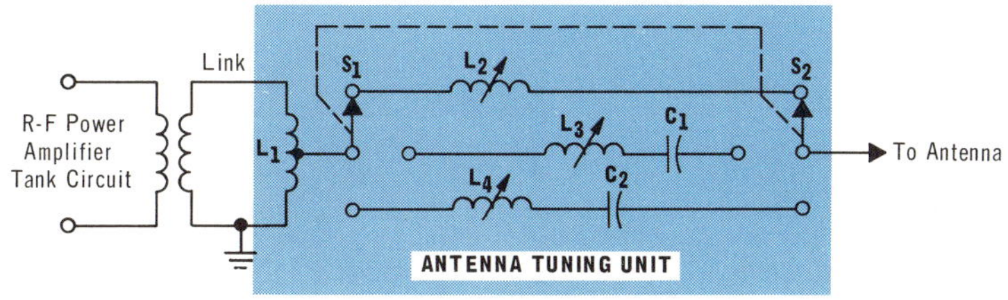

The reactance of a capacitive antenna would be balanced out by the inductive reactance of coil L_2

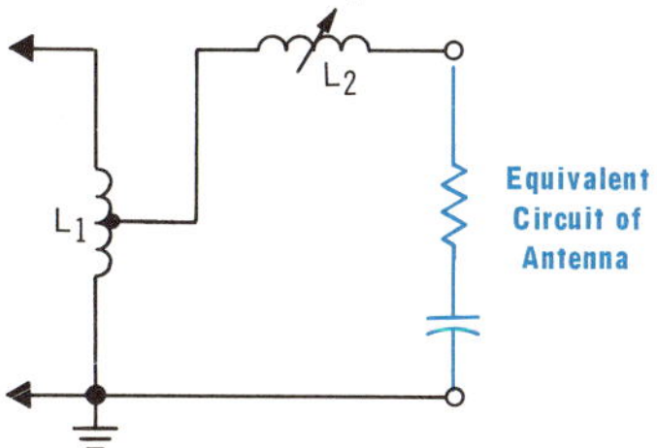

If the antenna contains both capacitive and inductive reactance, the antenna tuning unit introduces the appropriate components to balance them both out

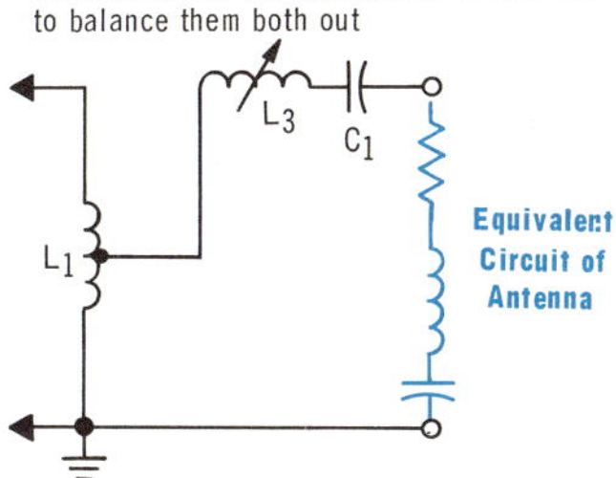

A basic antenna tuning unit is shown. In this circuit, switch S connects various capacitors and coils in *series* with the antenna circuit. These components are then adjusted to resonate with the reactive properties of the antenna. This makes the antenna a series resonant circuit, and, therefore, purely resistive.

The output of the r-f power amplifier is inductively (link) coupled to coil L_1 in the antenna tuning unit. L_1 has a variable tap that is set so that the resistance of the antenna is reflected back as the optimum load on the power amplifier tank circuit.

neutralization

You recall that when triodes are used as grounded-cathode r-f amplifiers, the reactance of the grid-to-plate interelectrode capacitance is often low enough to cause a significant amount of undesired feedback. This feedback is in phase with the input signal, and if it is not eliminated by neutralization, it can cause the stage to oscillate, thus producing unstable operation and distortion. In r-f tube-type voltage amplifiers, this problem is generally solved by using tubes or circuit configurations in which the unwanted feedback is not present. But in r-f power amplifiers, grounded-cathode triode stages are often desirable. When they are used, neutralization must be provided.

As you know, neutralization is merely the feeding back from *output* to input a voltage that is *equal* in *amplitude* to but *opposite* in *phase* with the undesired internal feedback through the tube interelectrode capacitance. The two voltages cancel each other and regeneration is eliminated.

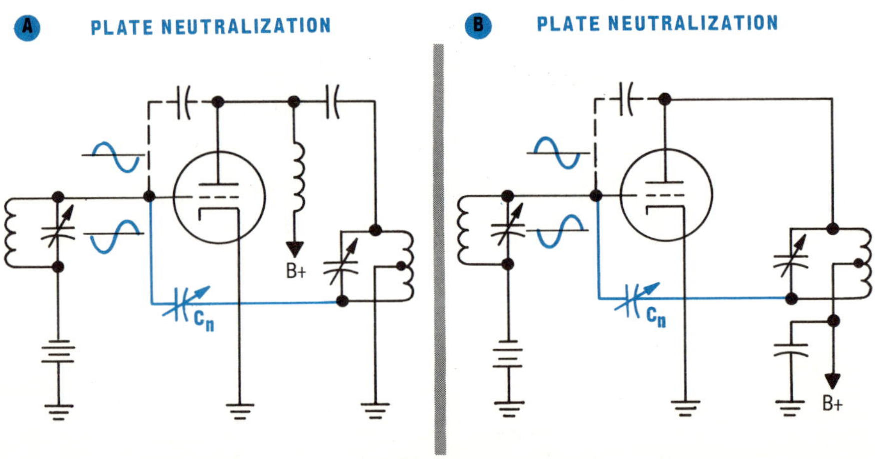

The voltage fed back to the grid through the external neutralization path (in color) cancels the voltage fed back through the grid-to-plate interelectrode capacitance (dashed lines)

The two basic methods of neutralization are called *plate neutralization* and *grid neutralization*. In plate neutralization, shown in A and B, the coil in the plate tank circuit is tapped to ground with respect to the r-f signal. This makes the r-f voltage at the bottom of the tank 180 degrees out of phase with the r-f signal at the top of the tank, again with respect to ground, and also, therefore, with respect to the grid. By connecting a capacitor (C_n) between the bottom of the tank and the grid, a voltage is fed back to the grid that is 180 degrees out of phase with the voltage fed back through the interelectrode capacitance.

neutralization (cont.)

C GRID NEUTRALIZATION

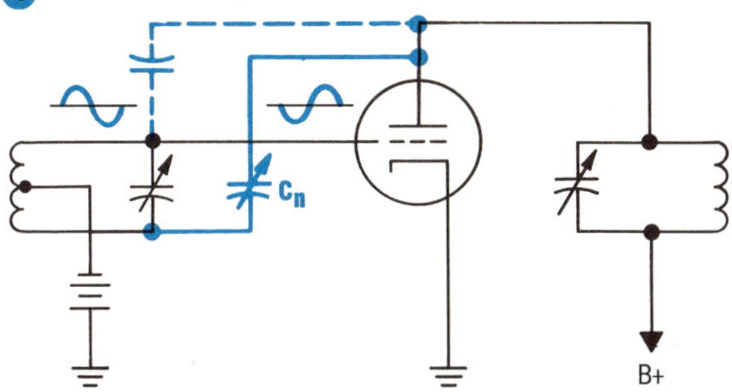

With inductive neutralization, inductor L_n forms a parallel resonant tank circuit with the interelectrode capacitance, and the resulting high impedance blocks the feedback. By making L_n adjustable, neutralization can be tuned to a number of different frequencies

D INDUCTIVE NEUTRALIZATION

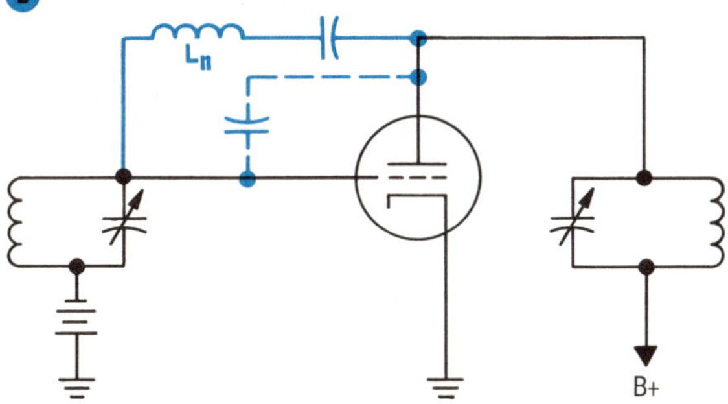

C_n is variable so that the neutralizing voltage can be made equal in amplitude to the undesired feedback voltage.

In grid neutralization, shown in C, the tapped coil is in the grid circuit. The unwanted feedback voltage is applied to the top of the tapped coil, and the neutralizing voltage is applied, through C_n, to the bottom of the coil. Since the coil is tapped, both voltages are again 180 degrees out of phase and cancel each other.

In a somewhat different method of neutralization, shown in D, a small coil (L_n) is connected between the plate and grid. The reactance of this coil is equal to the reactance of the grid-to-plate interelectrode capacitance at the signal frequency, so a parallel resonant circuit is formed between grid and plate. Such a resonant circuit has a high impedance at resonance, which prevents the undesired feedback voltage from being coupled back to the grid. The capacitor in series with the coil prevents the d-c plate voltage from reaching the grid. This method of neutralization is called *inductive neutralization* and is effective only at a *single* frequency, as determined by the reactance of the coil.

parasitic suppression

R-f power amplifiers are susceptible to what are called *parasitic oscillations*. These are oscillations that are *unrelated* and usually far removed from the frequency to which the amplifier is tuned. Parasitic oscillations can reduce the power output of the amplifier, cause excessive current that can shorten the life of the tube, and produce radiation of spurious signals that interferes with nearby electronic equipment.

The cause of parasitic oscillations is the development of *unwanted resonant circuits* that interact with each other in such a way that sufficient feedback is produced to sustain oscillation. These unwanted resonant circuits may be produced by the stray circuit capacitances interacting with the inductance of the circuit wiring, component leads, and tube elements, or they might be produced by a combination of these things and the actual circuit components themselves. In any case, the specific causes of parasitic oscillations vary from circuit to circuit. Because of this, the methods for eliminating parasitic oscillations, called parasitic suppression, also vary widely from circuit to circuit.

Generally, however, parasitic suppression involves the incorporation of small components in the circuit to: (1) detune one of the unwanted resonant circuits causing the parasitic oscillations; or (2) provide attenuation in the feedback path. Some of the circuits used for parasitic suppression are shown.

Small resistors or r-f chokes in series with the grid or plate connection provide parasitic suppression

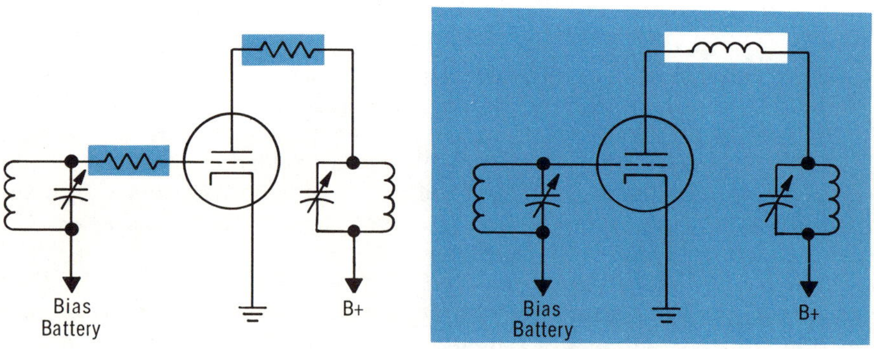

Resistors provide attenuation in the feedback path, while r-f chokes change the resonant frequency of one of the unwanted tuned circuits causing the parasitic oscillation

summary

☐ An r-f buffer amplifier is used to isolate, or act as a buffer for, one circuit from another. It gets its name from its function rather than its configuration. Ideally, a buffer amplifier is a class A, r-f voltage amplifier, although low-level class B or C amplifiers are sometimes used. ☐ In a radio transmitter, the buffer amplifier prevents loading of the oscillator by the succeeding amplifiers. ☐ Frequency multipliers are essentially class C r-f amplifiers. Frequency multipliers do not require neutralization because their input and output frequencies are different. ☐ R-f power amplifiers are used to increase the power level of an r-f signal. They operate either class B or C. ☐ For very high power requirements, tube-type r-f power amplifiers are used, but for low power applications, transistorized ones are being used increasingly.

☐ The last stage of a transmitter is the final power amplifier. ☐ In high-power AM transmitters, modulation takes place at the output of the final power amplifier. This allows all amplifier stages to be operated class C. ☐ In FM transmitters, modulation takes place before any of the power amplifier stages. ☐ Tuned circuits used with the r-f power amplifiers in AM transmitters generally have higher Q's and, therefore, greater selectivity than those in FM transmitters. ☐ For efficient power transfer, impedance-matching circuits must be used between cascaded r-f power amplifiers.

☐ Impedance-matching circuits must be used to match the output of an r-f power amplifier to any transmission line or antenna that it feeds. ☐ An antenna tuning unit is a separate unit that performs as an impedance-matching device between an r-f power amplifier and an antenna. ☐ Parasitic suppression is necessary in r-f power amplifiers to eliminate parasitic oscillations caused by unwanted resonant circuits.

review questions

1. What causes parasitic oscillations?
2. Why do frequency multipliers not require neutralization?
3. Why are buffer amplifiers used?
4. Why do FM r-f amplifiers require lower Q's than AM amplifiers?
5. What is an antenna tuning unit?
6. What are the two basic types of neutralization?
7. Are transistors ever used in power amplifiers?
8. Draw a frequency multiplier. Explain its operation.
9. In AM transmitters, modulation usually takes place in the output of the final power amplifier. Why?
10. Draw the circuit for a class C transistor amplifier.

i-f amplifiers

Intermediate-frequency (i-f) amplifiers are used in equipment such as superheterodyne receivers, where all input signals are converted to a single frequency (i-f frequency) for further processing after some degree of amplification. The i-f amplifiers provide the major portion of the *signal amplification* that takes place in an equipment and, at the same time, provide the necessary degree of *selectivity* so that adjacent-channel signals do not interfere with the desired signal. You will recall from Volume 2 that some equipments employ *double conversion,* in which two i-f frequencies and, therefore, two groups of i-f amplifiers are used. Essentially, though, the i-f amplifiers function in the same way. Only their frequencies are different.

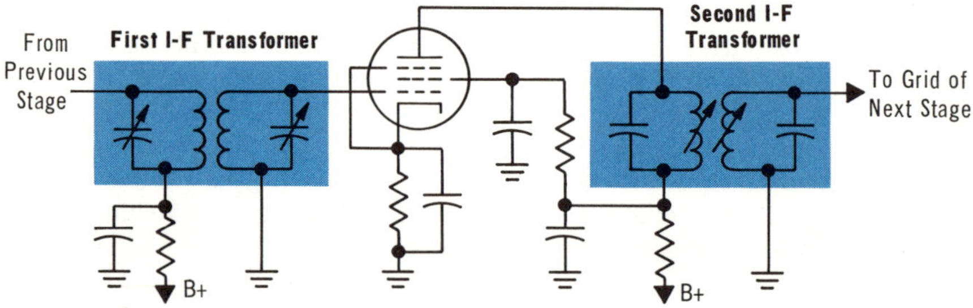

A basic tube-type i-f amplifier consists of a pentode class A voltage amplifier with double-tuned coupling circuits in both its input and output

An i-f amplifier is similar to an r-f voltage amplifier in that it is a tuned class A voltage amplifier. However, an r-f amplifier can be tuned to any frequency over its operating range, but an i-f amplifier is permanently tuned to a *single frequency.* Another difference is that r-f amplifiers rarely use *double-tuned coupling circuits* between stages, but with i-f amplifiers, double-tuned coupling is very common, especially with tube circuits.

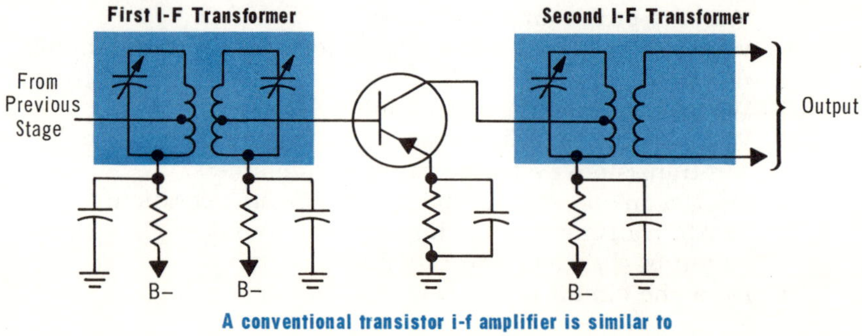

A conventional transistor i-f amplifier is similar to its tube counterpart in many respects

bandwidth

The *selectivity*, and, therefore, the *bandwidth* requirements of an i-f amplifier are somewhat contradictory in nature. A good degree of selectivity must be provided to prevent interference from adjacent channel signals. At the same time, however, the bandwidth must be sufficiently wide so that a signal and all its side-band frequencies are amplified equally.

The required bandwidth of an i-f amplifier depends on the frequency range of the side bands contained in the signal. This, in turn, depends on the type of signal, as well as on the degree of acceptable distortion. For example, an i-f amplifier for use in an AM broadcast receiver needs a bandwidth of 10 kilocycles, since such AM signals contain an upper and a lower side band, both 5 kilocycles wide.

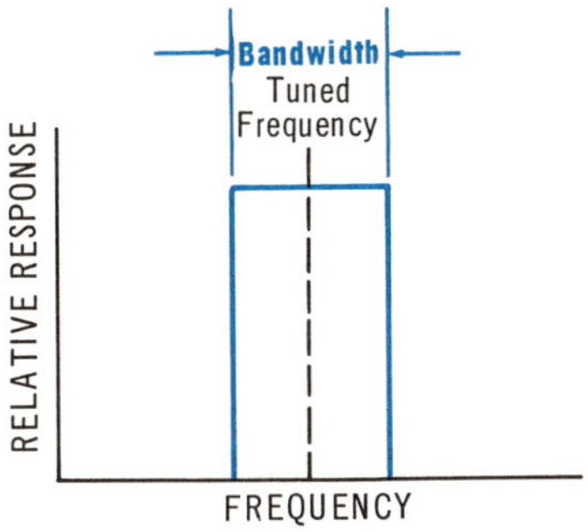

An ideal i-f amplifier would have a rectangular response curve

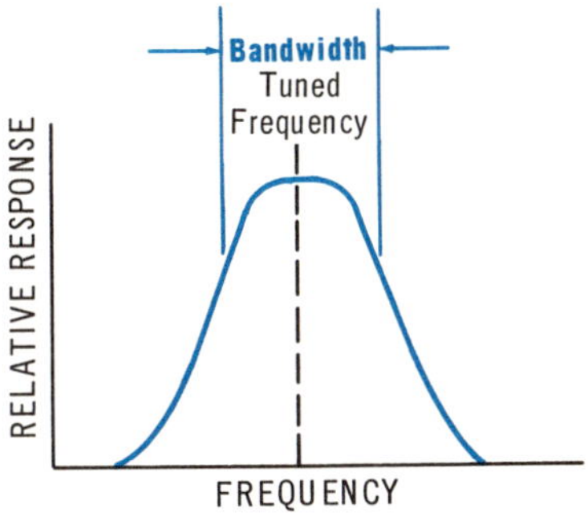

The response curve of a practical i-f amplifier has this basic shape

For some applications, such as voice communication, distortion of the signal is acceptable as long as it does not destroy the intelligibility. In these cases, the i-f amplifier could have a bandwidth as narrow as, say, 2500 cps. The signal would still be intelligible, and the increased selectivity would provide better discrimination against adjacent channel signals. The i-f amplifiers used for wide-band FM and television require relatively wide bandwidths since these signals contain a broad spectrum of side-band frequencies.

You can see from the above that although all i-f amplifiers perform essentially the same function, their selectivity and bandwidth requirements vary widely according to the type of equipment in which they are used. In all cases, though, their frequency-response curve is similar in shape to that shown.

The relatively flat peak of the curve indicates that all frequencies within the required bandwidth are amplified about equally. The steeply sloping sides indicate adequate discrimination against frequencies outside of the bandwidth.

effect of coupling on selectivity and bandwidth

In most electronic equipment, the necessary degree of selectivity cannot be achieved with a single i-f amplifier. You saw that selectivity in r-f voltage amplifiers was directly related to the *number* of tuned circuits used. The more tuned circuits, the greater the selectivity. The same thing is true for i-f amplifiers. For this reason, two or more i-f amplifier stages, each with tuned input and output circuits, are usually used for i-f amplification.

You will also recall from the discussion of r-f voltage amplifiers that the *shape* of the response curve of a single-tuned coupling circuit depends essentially only on the Q of the circuit, and the bandwidth of such a circuit is given by the equation: Bandwidth $= f_R/Q$, where f_R is the resonant frequency of the circuit. However, in double-tuned coupling circuits, such as those used with i-f amplifiers, the coupling circuit's *degree,* or coefficient, of coupling has a significant effect on selectivity, and, therefore, bandwidth. The degree of coupling is determined by the amount of *mutual inductance* between the primary and secondary of the coupling transformer, and this, in turn, depends essentially on the relative closeness of the two coils.

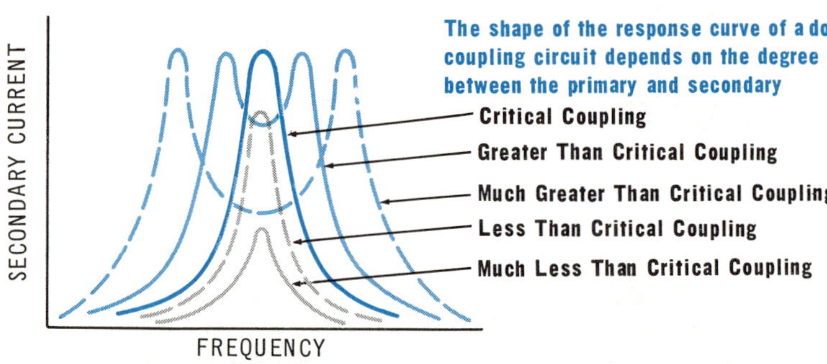

The shape of the response curve of a double-tuned coupling circuit depends on the degree of coupling between the primary and secondary

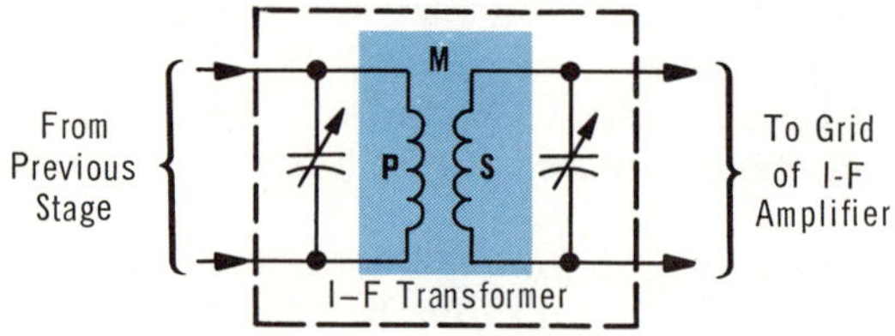

The degree of coupling is determined by the mutual inductance (M) between the primary and secondary coils

Coupling between tuned circuits is defined in terms of what is called *critical coupling*. This is the degree of coupling, as shown, at which the resonant current in the secondary circuit reaches its maximum value. If the coupling is less than critical (undercoupling), the secondary current has a lower value, and the circuit has a narrower bandwidth. When the coupling is greater than critical (overcoupling), the response curve widens (greater bandwidth), but displays two humps.

effect of coupling on selectivity and bandwidth (cont.)

You can see from the response curves that the various degrees of coupling possess both advantages and disadvantages as far as the selectivity and bandwidth requirements of an i-f amplifier are concerned. The curves for greater than critical coupling have steeply sloping sides, which means good frequency discrimination. They can also provide a relatively wide bandwidth. However, the humps indicate unequal response to difference frequencies within the bandwidth, and this is unacceptable since it would mean distortion of the signal. For critical and slightly less than critical coupling, good response within the bandwidth would be obtained, but the bandwidth is too narrow for most practical applications. Coupling much less than critical is generally undesirable because of both its narrow bandwidth and low response amplitude.

The frequency response of i-f amplifiers can be made to provide the necessary selectivity and bandwidth in one of two basic ways. One way is to use greater than critical coupling in the tuned coupling circuits, and then modify the circuit to produce a more linear response over the bandwidth. As shown in A, this can be done by placing *resistors* in *parallel* with the L and C components of the tuned circuits. These shunt resistors have the effect of adding resistance in series with the coils and thus lower the Q of the tuned circuits. The result is a flatter response, although at some sacrifice in amplitude.

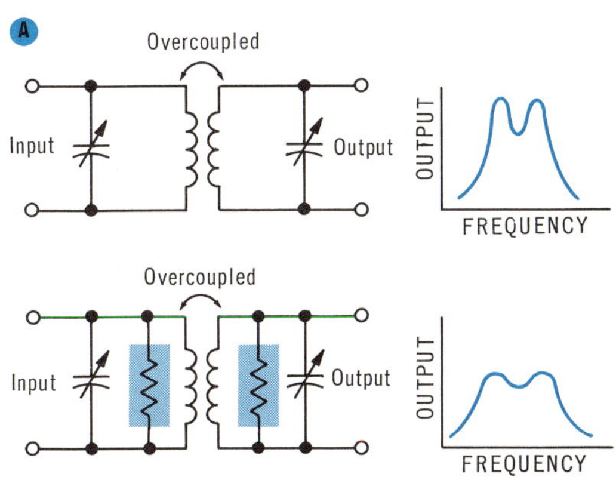

By adding shunt resistors across the tuned circuits, the dip in the overcoupled frequency response curve can be smoothed out. Some loss in gain results, however

Another method of making the response more linear (shown in B) is to *overcouple* one i-f transformer to obtain the required bandwidth and then use another i-f transformer with a *single-peak* response curve to "fill in" the dip in the response. The single-peak response can be provided by either a single-tuned coupling circuit, or a double-tuned circuit with critical or slightly less than critical coupling.

practical i-f coupling circuits

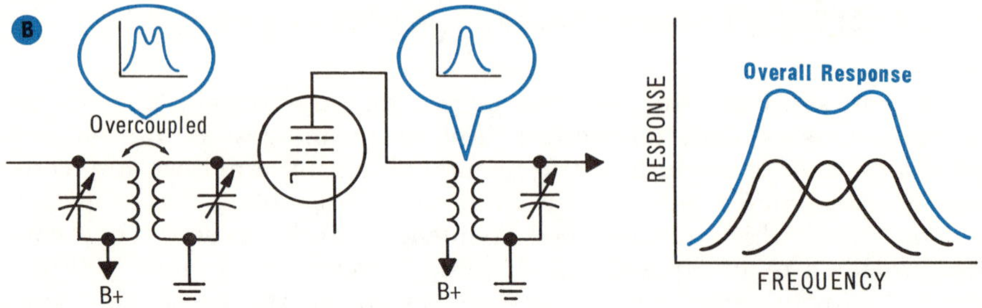

The desired response can be obtained by using a coupling circuit with a single-peak response to fill in the dip produced by an overcoupled circuit

The other common method of obtaining the required response is to use different tuned circuits that are tuned to slightly *different* frequencies. The circuits then all have single-peak response curves slightly displaced in frequency from each other. Since the overall response is the product of the individual responses, both the required bandwidth and an essentially flat response are obtained. This method is called *stagger tuning* and is illustrated in C. While stagger tuning can also be used with transformer coupling, it is more frequently used with LC coupling circuits. This is known as *impedance coupling,* where each coil resonates with its interwinding capacitances, and a capacitor couples the signal to the next stage. The circuit shown uses six tuned coils, with each pair tuned to one frequency. If all six were tuned to different frequencies, an even flatter overall response would be obtained.

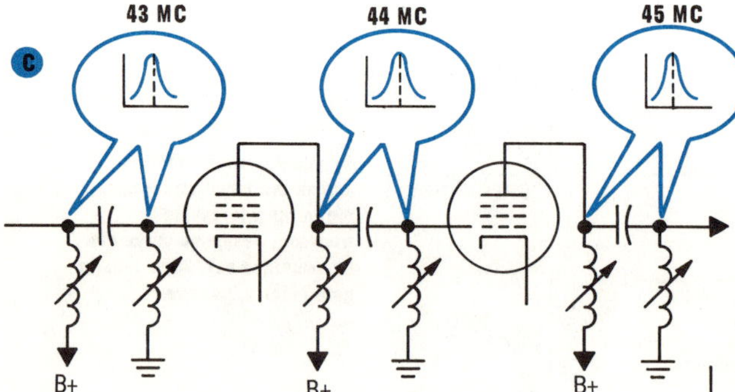

In stagger tuning, the resonant circuits are tuned to slightly different frequencies

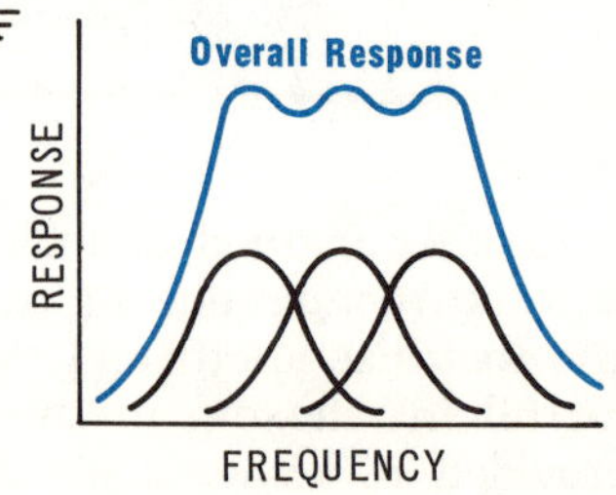

The individual response curves combine to produce the overall response. Although single-tuned impedance-coupled circuits are shown, transformers with less than critical coupling can also be used

frequency response of television i-f amplifiers

So far, all of the i-f frequency response curves shown have been *symmetrical* about the tuned frequency. This is a requirement for i-f amplifiers that must process *double side-band* signals. Since the side bands on either side of the carrier of a double side-band signal contain equal intelligence, both side bands must be amplified equally, or else the signal will be distorted.

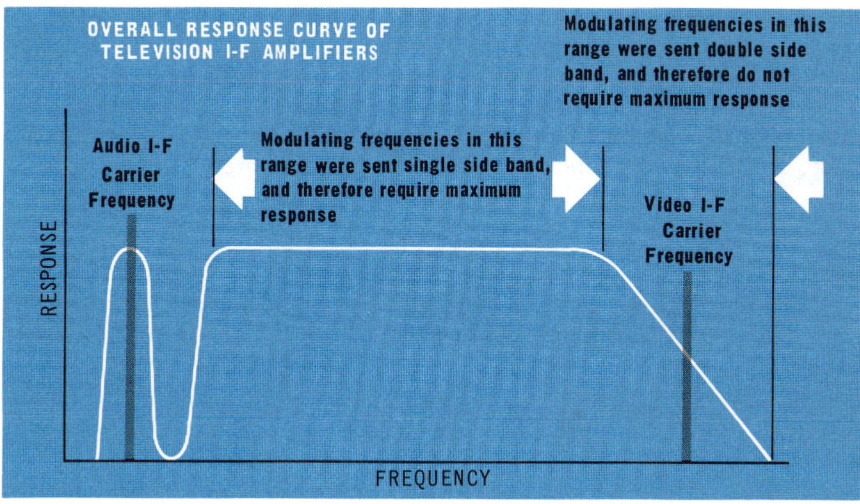

In i-f amplifiers used in television receivers, though, the situation is different. You recall from Volume 1 that *vestigial side-band* AM transmission is used for the video portion of the television signal. All frequencies in the lower side band more than 0.75 megacycles from the carrier are rejected and not transmitted. This means that modulating frequencies greater than about 0.75 megacycles are transmitted as single side-band signals, and those less than 0.75 are transmitted as double side-band signals. Since the modulation energy in both side bands is ultimately combined to reproduce the original intelligence, a symmetrical response in the receiver i-f section would result in a demodulated signal output twice as great for video signals below 0.75 megacycles as for signals above 0.75 megacycles. This would cause an unacceptable degree of distortion.

To compensate for the lower amplitudes of the video modulation frequencies above 0.75 megacycles, the response of the i-f amplifiers is not symmetrical about the video carrier frequency. Instead, the response is adjusted so that frequencies above 0.75 megacycles are amplified more than those below 0.75 megacycles, with the result that the relative amplitudes of all the original modulating frequencies are restored.

filter selectivity

Although tuned coupling circuits are the most common way of providing i-f amplifier selectivity, in some equipments practically all of the i-f selectivity is provided by a single *selective filter* at the input to the first i-f amplifier. One of the most frequently used selective filters is a *piezoelectric crystal.* Such a crystal is essentially a *series* LCR circuit, with its thickness, shape, and material determining the resonant frequency. In the circuit shown, a crystal filter is connected between two conventional tuned circuits in the input to the first i-f amplifier. The input i-f signal, usually from a mixer stage, is transformer-coupled to the parallel resonant circuit consisting of coil L_1 and capacitor C_1. To frequencies within its resonant bandwidth, the crystal offers a very low resistance, and these signal frequencies are coupled through capacitor C_3 to the tuned circuit of the first i-f amplifier (L_2C_4). To frequencies outside of its resonant bandwidth, the crystal offers a very high resistance, and these frequencies are effectively attenuated.

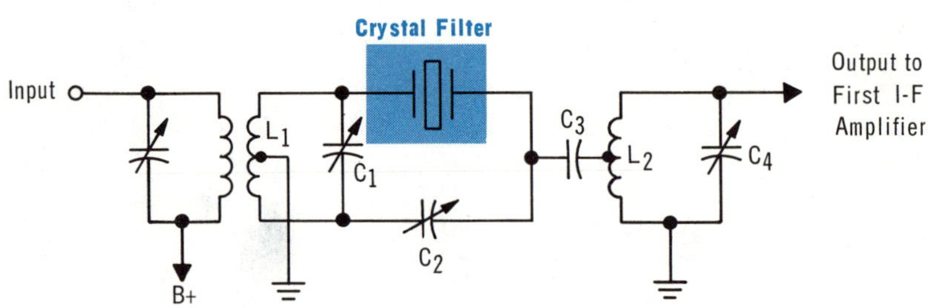

Besides crystal filters, other mechanical-type filters are sometimes used with i-f amplifiers to provide the required selectivity and bandwidth

Capacitor C_2 is called the crystal-phasing capacitor. It is adjustable and is used to neutralize the capacitance that exists between the crystal and its holder. Coil L_2 is tapped to provide an impedance match between the crystal and the input circuit of the first i-f amplifier.

By using a crystal filter for i-f selectivity, the requirements of the coupling circuits between i-f amplifiers are greatly reduced. Basically, all these circuits have to do is preserve the frequency response established by the crystal filter. Sometimes, tuned coupling is even eliminated, and adequate RC or LC coupling can be used.

In many communications receivers, a crystal filter can be switched in and out of the circuit. For reception of CW signals, the crystal is connected in the i-f circuits and provides a very narrow i-f bandwidth (as narrow as 50 to 100 cps) and sharp discrimination against frequencies outside of this band. For modulated signals, such as voice, the crystal is switched out of the circuit, and the i-f selectivity and bandwidth are provided by conventional tuned coupling circuits.

gain

Normally, the gain of an i-f amplifier cannot be adjusted manually. However, in many applications, such as radio and television receivers, the i-f gain is automatically adjusted to compensate for changes in the strength of the received signal

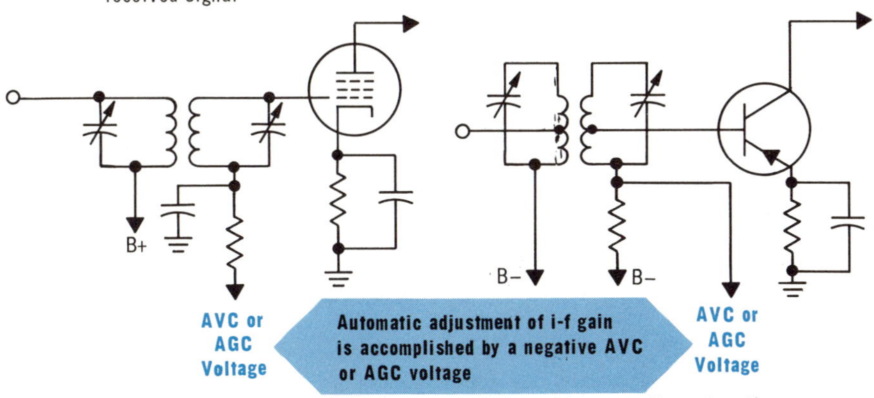

This automatic adjustment of gain is accomplished by a negative automatic volume control (AVC) or automatic gain control (AGC) voltage produced at a later stage in the equipment. The AVC or AGC voltage varies the grid voltage, and, therefore, the stage bias in tube i-f amplifiers, and usually the base voltage, and, therefore, the base-emitter bias in transistor amplifiers

As was pointed out previously, i-f amplifiers provide by far the largest portion of the overall signal gain in equipments in which they are used. Generally, their inputs are in microvolts or millivolts, and they must deliver an output signal of a few volts to a detector stage. To supply this gain, it is usually necessary to connect two or more i-f amplifier stages in cascade. In tube circuits, pentodes, with their high gain, are most commonly used for i-f amplifiers. Transistors, because of their lower gain, normally require more cascaded stages than do tubes to provide the required overall gain. In some tube circuits, even though a certain number of stages can supply the required gain, they cannot provide the necessary selectivity and bandwidth. Additional stages and their accompanying tuned circuits are then used for *response purposes,* with their gain being of secondary importance.

The high gains of i-f amplifiers increase the chance of *stray coupling* between stages when a cascade arrangement is used. The greater the number of cascaded stages, the more chance there is of feedback through the stray coupling. This can cause instability and oscillation, and result in distortion and a decrease in gain. As a result, careful shielding, good circuit layout, and effective *decoupling* networks are important in cascaded i-f amplifiers. This problem is somewhat lessened when *double conversion* is used, since all of the i-f stages do not then operate at the same frequency.

neutralization

Normally, tube-type i-f amplifiers do not require neutralization, since they use *pentodes*. However, at FM and higher frequencies, high-gain pentode i-f amplifiers have a tendency to oscillate because of feedback through the tube interelectrode capacitance. These capacitances are very small, but the combination of *large gain* and *high frequency* can cause sufficient feedback to cause oscillation.

When such conditions exist, conventional triode neutralizing circuits can be used, but the small values of capacitance involved make such neutralizing circuits hard to adjust. A different method of neutralizing a pentode stage is to use a *common* bypass capacitor for the plate and screen grid circuits. Such a capacitor effectively forms a *bridge circuit* with the tube interelectrode capacitances and cancels any regenerative feedback that would otherwise reach the grid. Parasitic suppressor resistors can also be used in the control grid, screen grid, or plate circuit, as with r-f amplifiers.

Most transistor i-f amplifiers are neutralized to eliminate partially the internal collector-to-base feedback. The circuits used are similar to those employed for the neutralization of r-f transistor amplifiers. However, the neutralization requirements for i-f stages are less severe because of the lower frequencies involved and the fact that an i-f amplifier always operates at the same frequency. Frequently, transistor i-f amplifiers are designed so that they do not require neutralization. Almost always, though, this is done at some sacrifice in gain. Methods that can be used for this include an *intentional mismatch* between the outputs and inputs of succeeding cascaded stages and the use of *shunt damping resistors* in collector-tank circuits.

Often, oscillation takes place in i-f sections because the signal from one of the later stages is fed back to an earlier stage through the B+ line. To prevent this, the plate circuit of each i-f stage is filtered by an *RC decoupling network* to keep the i-f signal out of the power supply.

C_4 is a common bypass capacitor for the plate and screen grid circuits. As shown at the right, it forms a bridge circuit with the tube interelectrode capacitances. Any feedback voltages that would otherwise be coupled back to the grid (1) from the plate (2) and the screen grid (3) are canceled by the lower arms of the bridge (C_3 and C_4)

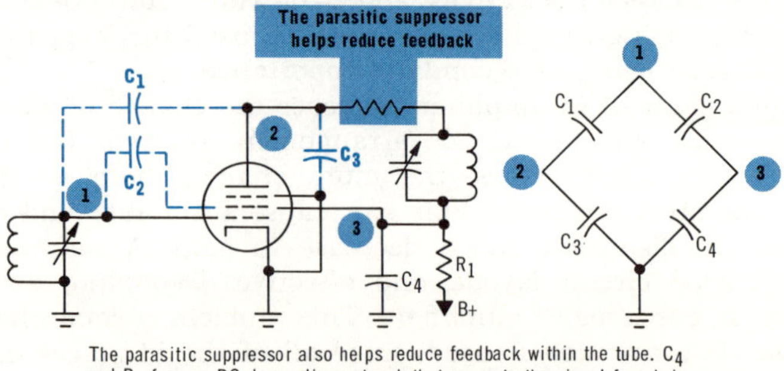

The parasitic suppressor also helps reduce feedback within the tube. C_4 and R_1 form an RC decoupling network that prevents the signal from being fed through the B+ line to another stage

typical i-f amplifiers

A two-stage, transistor i-f amplifier of the type used in FM receivers is shown in A. The required FM bandwidth is obtained by using double-tuned coupling circuits that are slightly overcoupled. The dip in the response curve caused by overcoupling is eliminated by providing a degree of mismatch between the tuned coupling circuits and the transistor inputs and outputs. This mismatch dampens or lowers the Q of the tuned circuits, thereby flattening out their response. The primary of each i-f transformer is mismatched to its respective transistor output by being connected directly in the collector circuit. The secondary of each transistor is mismatched to its transistor input by means of a capacitive voltage divider.

A

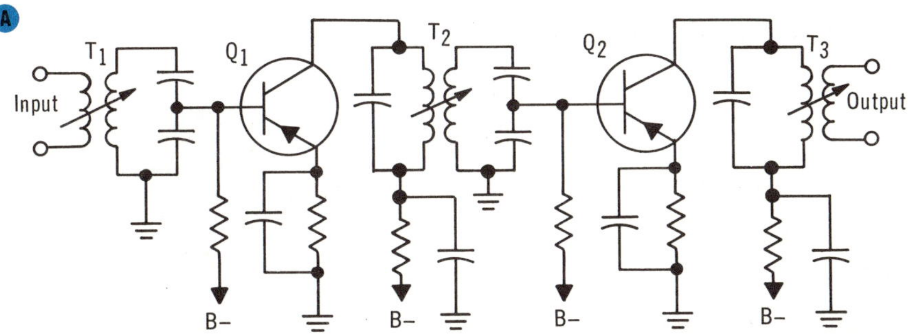

Very often, the individual stages of cascaded i-f amplifiers are identical. A careful analysis of the operation of one stage can provide a good understanding of the operation of the complete i-f section

B

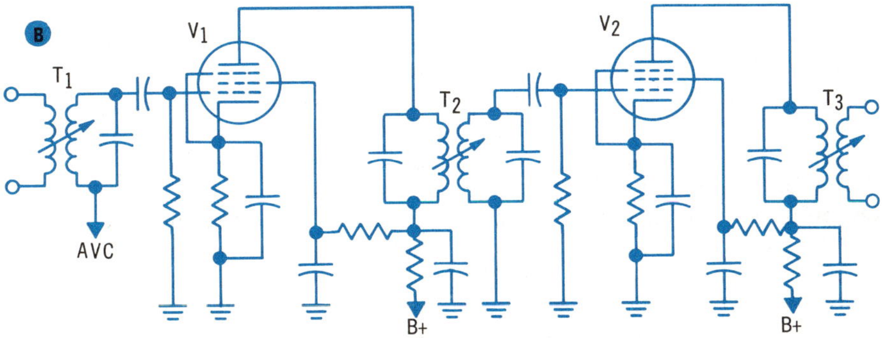

A typical two-stage, tube-type i-f amplifier is shown in B. Again, double-tuned coupling circuits are used throughout. In this amplifier, though, the necessary bandwidth is achieved by stagger-tuning i-f transformers T_1, T_2, and T_3. Aside from the coupling circuits, each stage operates as a conventional class A voltage amplifier. Plate decoupling networks keep the i-f signal out of the B+ line.

i-f limiters

I-f limiters are used in *FM* receivers between the last i-f stage and the detector. The purpose of an i-f limiter is to deliver to the detector a *constant-amplitude* i-f signal. This eliminates any amplitude variations that have been imposed on the signal by noise or other interference. If these amplitude variations are passed on to certain types of detectors, they will appear in the demodulated signal as distortion or as annoying sounds.

The type of detector used generally determines whether a limiter is necessary. Some types of detectors, such as the Foster-Seeley discriminator, are sensitive to amplitude variations in their inputs, and therefore *must* be preceded by a limiter. Other detectors, such as the ratio detector and the gated-beam detector, do not readily respond to amplitude variations in their inputs. So, limiters are not required when these detectors are used, but are sometimes used to ensure good results.

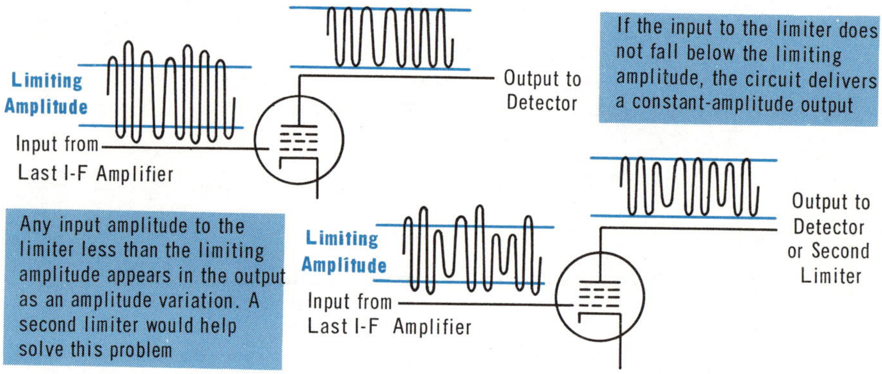

Essentially, a limiter removes, or clips, both the positive and negative *peaks* of the signal whenever they exceed a certain amplitude, which we will call the *limiting amplitude*. Thus, if the input to the limiter never falls below the limiting amplitude, the clipping action of the limiter will produce a constant-amplitude output. However, if the input signal should drop below the limiting amplitude, a corresponding amplitude variation would appear in the output and be passed to the detector. In other words, the limiter is *ineffective* against amplitude variations *below* the limiting amplitude. This fact constitutes one of the basic requirements of any limiter. That is, that the signal must be amplified sufficiently by the i-f stages so that when it reaches the limiter its minimum amplitude is equal to or greater than the limiting amplitude. In some receivers, where there might be a problem, more than one limiter is used in cascade, so the second limiter can take care of what the first one missed.

The two basic types of limiters are explained on the following pages. Quite often, both of these techniques are used in one stage.

the plate limiter circuit

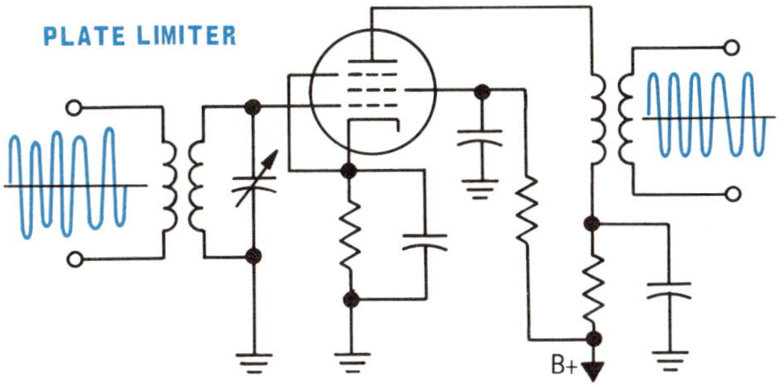

The plate limiter achieves its limiting action by virtue of its low screen and plate voltages

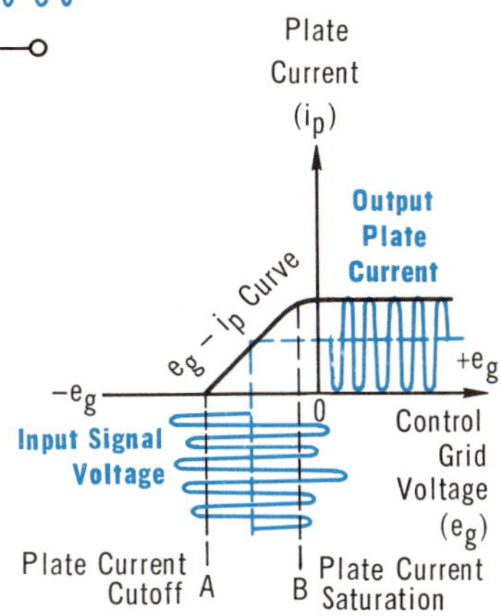

Signal peaks more negative than A are clipped as a result of plate current cutoff, while peaks more positive than B are clipped through plate current saturation

One basic type of limiter is the plate limiter. The bias and operating voltages for the stage are set so that the positive and negative values of the limiting amplitude correspond to *plate current saturation* and *plate current cutoff*, respectively. When a peak of the input signal exceeds the *positive* limiting amplitude, the grid voltage causes plate current saturation. No additional *plate* current can flow no matter how much further the signal swings positive, so the output signal is clipped at the limiting value. When a peak of the output signal exceeds the *negative* limiting value, it is sufficiently negative to cut off the *plate* current, so again the output is clipped at the limiting value.

To achieve this limiting action, the stage is operated with very low screen grid and plate voltages. The low screen and plate voltages cause the positive signal peaks to drive the tube into saturation very easily. Because of the low plate and screen voltages used, the bias must also be low to allow current to flow. However, because of the low voltages, very little additional bias is needed to cut off the tube. The negative swings of the signal do this easily.

In tube circuits, *sharp cutoff* pentodes are normally used for plate limiters, because there is a small range between cutoff and saturation. Transistor versions of the plate limiter operate in essentially the same manner as described above. The transistor bias is set so that excessive amplitude peaks are clipped in the collector output.

Because of its low screen and plate voltages, which may be 30 volts or less, a plate limiter produces very little gain of the i-f signal. In some applications, this is a disadvantage. Where gain as well as limiting are desired, the *grid limiter* can be used. This is described on the following page.

the grid limiter circuit

The grid limiter circuit uses higher operating voltages than the plate limiter, so it can provide gain as well as limiting. The typical grid limiter shown in A operates essentially as a class C amplifier using *grid-leak bias.* The bias is developed by capacitor C_1 charging when the grid draws current on the positive peaks of the input signal and then discharging through resistor R_1 when the signal amplitude drops to cause negative voltage drop, which serves as the bias. As shown in B, the larger the amplitude of the input signal, the farther the stage is biased beyond cutoff.

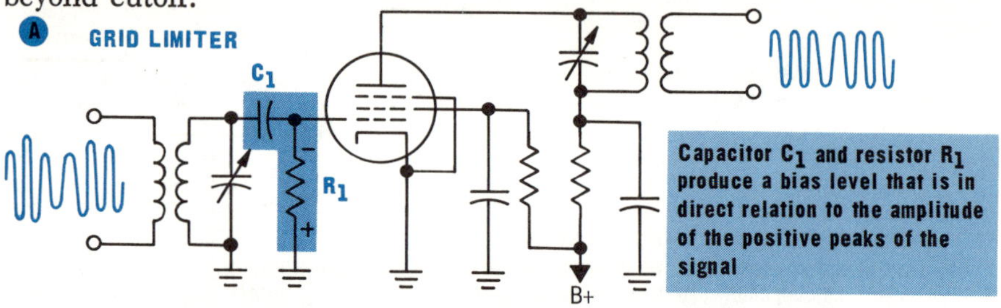

Because of the action of C_1 and R_1, the input signal, regardless of its amplitude, produces an equal pulse of plate current. The plate tank circuit then converts these pulses to a constant-amplitude FM output

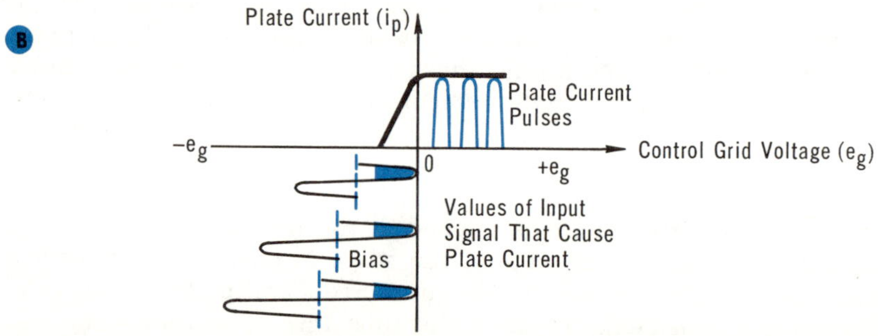

Since the shift in the bias point is directly proportional to signal amplitude, each cycle of the input signal will cause plate current to flow for an approximately *equal* period of time. Thus, the output voltage has a constant amplitude.

The *RC time constant* must be long with respect to the time of one cycle of the i-f frequency. This is necessary to maintain the bias constant as long as the input signal amplitude is constant. At the same time, the time constant should be short compared with one cycle of the lowest frequency amplitude variations that are to be clipped. If it is not, the bias will be determined by the lower-frequency amplitude variations, and the higher-frequency variations will not be limited at all.

When various types of AM interference must be removed from a signal by limiting, sometimes two limiters are connected in cascade, and each has a different time constant for its grid-leak bias circuit.

summary

□ The required bandwidth of an i-f amplifier depends on the frequency range of the side bands contained in the signal. □ With double-tuned coupling, which is common in i-f amplifiers, the degree of coupling has a significant effect on selectivity. □ Undercoupling results in a narrower bandwidth than does critical coupling, while overcoupling produces a wider bandwidth, but with two humps in the response curve. □ Stagger tuning provides a very flat frequency response with a large bandwidth.

□ Television i-f amplifiers have nonsymmetrical frequency-response curves to compensate for the vestigial side-band transmission used for the video signal. □ Crystal filters, which operate on the piezoelectric effect, are very selective, and are sometimes used to provide all of the i-f selectivity in an equipment. □ Because of the high gain of cascaded i-f amplifiers, effective decoupling networks are important to prevent instability and oscillation.

□ At lower frequencies, i-f amplifiers do not require neutralization because they use pentodes. However, at FM and higher frequencies, high-gain i-f amplifiers tend to oscillate as a result of feedback, and neutralization is required. □ Most transistor i-f amplifiers are neutralized to partially cancel internal collector-to-base feedback. □ I-f limiters clip the amplitude extremes of a signal. □ Plate limiters clip by means of plate current saturation and plate current cutoff. □ Grid limiters clip by means of an automatically varying grid bias.

review questions

1. What is *stagger tuning*?
2. Where gain, in addition to limiting action, is desired, should a plate or grid limiter be used?
3. Must tube-type i-f amplifiers be neutralized?
4. Why do television i-f amplifiers have unsymmetrical response curves?
5. What is meant by *critical coupling*? *Overcoupling*? *Undercoupling*?
6. For each type of coupling, draw a typical response curve.
7. In cascaded i-f amplifiers, why is effective decoupling important?
8. What is *double conversion*? Why is it used?
9. What is the equivalent circuit of a crystal filter?
10. Draw a grid limiter circuit. Explain its operation from a waveform standpoint.

index